THE OWNER'S MANUAL FOR THE HUMAN BODY™

2001 MANUAL

(A WORK IN PROGRESS)

JAMES P. FRACKELTON, M.D.

Cover Paintings: Darline S. DeStephen, Ashland, Virginia
Cover Design: Ryan Feasel
Graphs: Douglas MacTaggart, Ph.D.
Art: Ryan Feasel

DEDICATION

To my wife, Polly, whose encouragement to write the manual was unfaltering. I will always be grateful for her help in bringing the book to fruition.

DISCLAIMER

This book is intended to be a guide for improving your general knowledge. It is not intended to be used for medical diagnosis or medical treatment. This book is in no way intended to be a substitute for health care by a licensed health care professional. Individual health problems must be treated and closely supervised by a licensed physician. Before changing any therapies consult with your physician of choice. Do not hesitate to seek second opinions.

ACKNOWLEDGEMENT

Thanks to Melissa Szabo and Cristie Gardner for their wonderful help in typing and composition, and Maria Patrick for her dedication to Apple Press as a resource for all of us.

CONTENTS

ABOUT THE AUTHOR

James P. Frackelton, M.D., graduated from Yale in 1949 and in 1954 from Western Reserve Medical School (now Case Western Reserve University).

He served in the Navy twice, the second time as a Naval Flight Surgeon. He was founder and President of the Medical Service Foundation of America during the 1960s. He is a past chairman of the Family Practice Department Fairview General Hospital in Cleveland, Ohio. He established the Preventive Medicine Group in Cleveland in 1976. He is past Vice Chairman of the American Board of Chelation Therapy, past President of the American College for Advancement in Medicine, and current President of the American Institute of Medical Preventics. He has lectured to physicians around the world on Preventive Medicine and Chelation Therapy, including 400 physicians in Beijing, China, in 1997.

Dr. Frackelton is presently practicing full time Preventive/ Integrative Medicine in Cleveland with the Preventive Medicine Group.

PREFACE

It is presumptuous for anyone to write The Owner's Manual for the Human Body. However, I believe that it is important for everyone to have an overall view of how our wonderful self-repairing body works. This manual is written to be a continuing work in process with updates of information in future issues. The understanding of some basic principles of how the system works will allow you to better evaluate events in your life and perhaps troubleshoot the problems in a more "scientific" manner. We are bombarded continuously with conflicting information about health and disease in a manner that can be confusing. If you wish to achieve good health, take the initiative to start caring for yourself and to work in partnership with the physician you choose.

We have a serious problem in modern medicine, which comes about by the "extinction principle" in specialization of medicine and surgery. As you learn more and more about less and less you learn everything about nothing.

We certainly need specialization in medicine. You would want to have the finest eye or brain surgeon if the need arose. However, it is critical for these physicians and surgeons to understand how various self-repair systems operate and maximize their impact. In general no specialist does this.

Because physicians "fix" problems, the average physician knows a great deal about his or her expertise in disease, but unfortunately knows very little about the true principles of health. A Health Maintenance Organization (HMO) is an oxymoron run by "financial experts;" hence, our health industry does not give us what we think we are paying for.

This manual is designed to give you a tree of many branches from which you can organize the details of health information you will glean from the mass of books and articles available under the general term of health (or lack of it).

By the nature of the complexity of our bodies only general concepts will be presented. Some of the information you may already know, and some will be new. These concepts will help you to evaluate media information and separate truth from information promoted with an economic agenda from public relation "fronts."

People empower standard medicine to be "crisis oriented." Drive the car until all of the warning lights flash on the dashboard, then come in and we will replace the engine. In the United States we spend the most money for "health care," but we do not have the best health. Perhaps if the emphasis goes more to prevention we can finally have what we want in improved quality and quantity of life. The impetus for the work in prevention starts and continues within your power.

There are many natural compounds that have useful effects and sometimes profound effects on the improvement of our health. However, natural compounds cannot be patented. Interest is low in proving the efficacy of these natural substances. Profit is not assured without a patent.

However, if a drug company can change the complex molecular structure just enough to be patentable (not found in nature), the company can proceed to patent the new compound. With permission from FDA they will invest the money to set up studies to prove the efficacy of this new compound in the same realm as was the original natural compound. Too often the new compound will not be anywhere near as effective as the natural compound.

Basic research in natural compounds and other areas of alternative medicine need to be funded by non profit-seeking foundations.

Most drug and surgery scientific studies are not based on science. A double blind placebo controlled study of a drug generally has limited bearing on how the body's system works; otherwise, the cause of the disease in the study would be a lack of the drug in the study. The rapid growth of alternative medicine is due to the fact that the principles are based more on basic research than are most principles of standard medicine. Unfortunately, lack of funding for good research has made a lot of alternative medicine claims suspect. As you know, the patent exclusivity pays for most of the drug industry's research and development. Because of competition, standard medicine disregards and condemns alternative therapies due to "lack of proof." It has been estimated that only about 15% of what the standard physician does has been scientifically proven by basic research.

You are invited to recommend additions for future issues.

INTRODUCTION

A crime has been committed. The crime is the loss of your good health. You want the crime solved! You need a "Sherlock Holmes" investigator to solve the crime. After careful investigation it appears that the crime (to your body) resulted in illness that was a failure of your self-repairing systems. You did it!

The goal of this manual is to enable you and your doctor to develop ways to return your stolen health by getting your self-repairing systems up, running, and finely tuned.

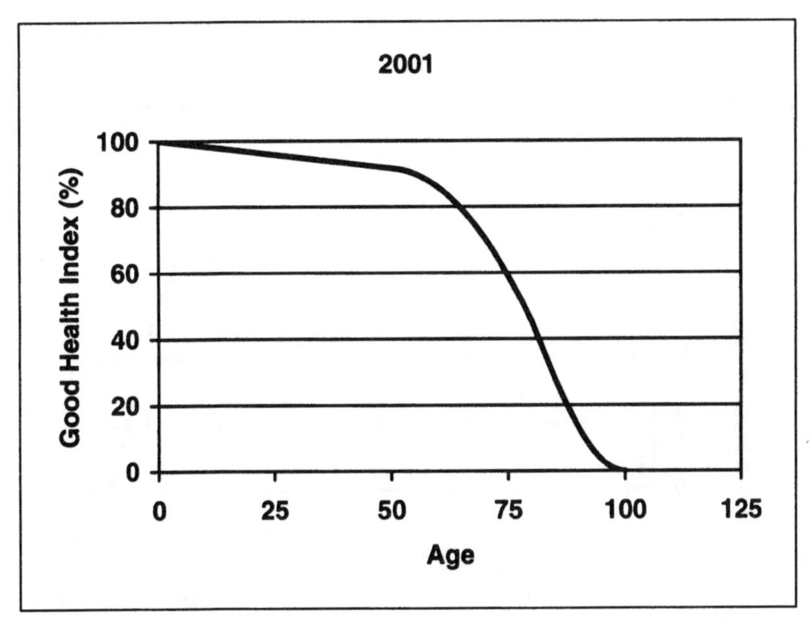

Figure 1

In the beginning of the twentieth century (1900), the life expectancy was about 45 years. In 2000 life expectancy has increased to about 76 years (79 years for women and 73 1/2 for men). This increase occurred primarily because of the delivery of clean water and the advent of antibiotics. This increase in life expectancy sounds great, but is it really what it appears to be? If you were 75 years old in 1900 you had an average of 7 more years of life than expected. If you were 75 years old in 2000 you have an average of 10 more years of life than expected. Shouldn't the miracles of modern health care, modern medicine and the availability of modern food do better? With the huge increase in cost of health care—are we really gaining a healthier, longer life? If not, why not?

Unpatentable natural compounds need much greater basic research funding despite little chance for financial return on their investment. Natural compounds are much more likely to have an effect on "boxing the Health Index vs. Age Graph" because they are part of the natural system from which we were all created.

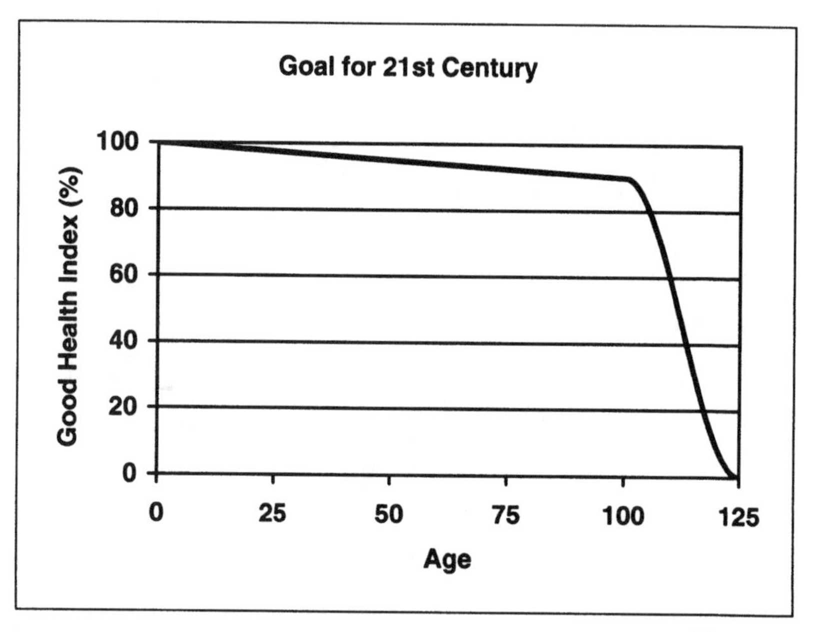

Figure 2

I. HOW THE SYSTEM OPERATES
The Cell

A single principle of good health is to optimize cell energy. Give the body what it needs. Take away what the body doesn't need.

Good health may appear simple but it is exceedingly complicated to achieve. The medical profession is just beginning to understand how the system works. Scientists can spend their entire lifetime studying one particular enzyme in cellular metabolism but not really know all about it.

The average 170 pound body has been estimated to be composed of 100 trillion cells (100 million million) which have developed from the female cell fertilized by the male sperm to become a super cell; the fertilized egg. Upon the exponential division of this cell and the preprogrammed genetics of embryology, various new cells become the brain, heart, skin, muscles, bones, etc.

Whatever is good for the cells is good for the person and vice versa. Our health and subsequent longevity is dependent on cell energy. We will discuss some of the cellular functions and how injury can reduce this cell energy.

The cell wall can be injured by trauma and more frequently by free radical damage from chemicals or irradiation. A free radical is a reactive oxidative chemical species which is damaging to the vulnerable unsaturated fatty acids making up the cell membranes. A protein called Super Oxide Dismutase (SOD) is present in the cell and attempts to prevent this damage from free radicals. SOD is the fifth most common protein in the body and is very important.

In different animal systems there is a direct correlation between the animal's longevity and the sophistication of the SOD antioxidant system. The internal liquid of the cell (cytosol) has a SOD which contains zinc and copper. The mitochondrial

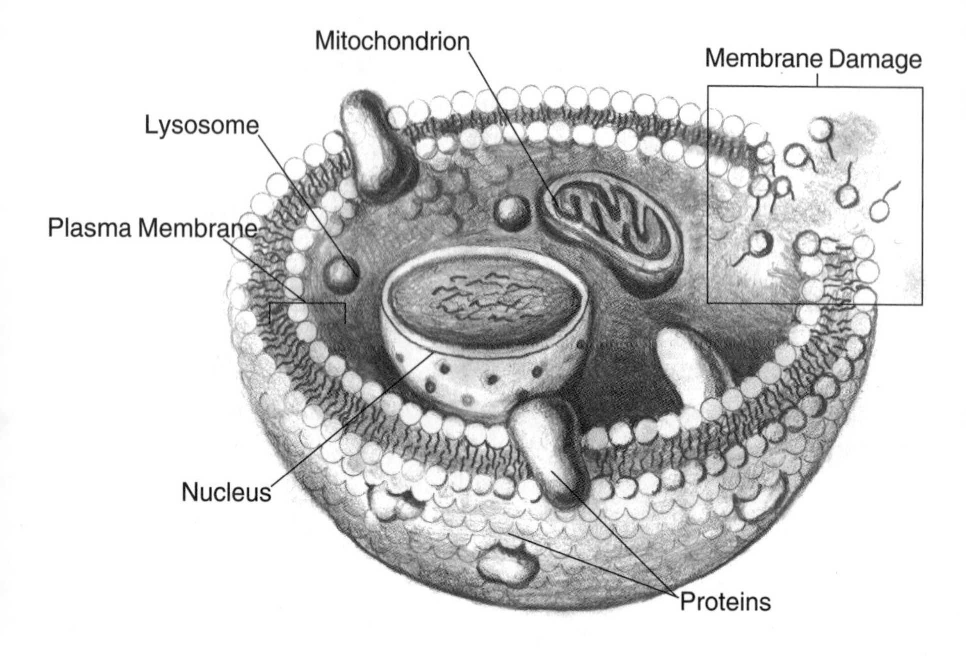

powerplant is protected by manganese SOD. The mitochondria produce the energy called ATP (Adenosine Triphosphate) with which all cells operate. It has been estimated that mitochondria produce about 100 pounds of ATP each day (ATP vibrates back and forth with ADP, or Adenosine Diphosphate).

A good percentage of the action of the cell takes place in the cell membrane. If cells didn't communicate with each other and work in a coordinated manner we would just be a sac of 100 trillion cells. There is an electric voltage potential maintained between the inside and the outside of the cell membrane. This potential is discharged during communication between cells. The potential is recharged through the action of magnesium ATP dependent pumps (pumping out the sodium and returning the potassium).

In a similar movement, calcium ions entering the muscle cell through "calcium channels" trigger the muscle cell contractions. Magnesium plus ATP from the mitochondria operate the pump that moves calcium out of the cell which then triggers the relaxation of the muscle cell.

Why is this discussion important? The loss of energy of the cell is caused by the extracellular minerals sodium and calcium moving in excess into the cell, or the loss of intracellular minerals magnesium and potassium through a damaged cell membrane.

When calcium enters the cell wall in excess it goes to the mitochondria and combines with phosphorus to produce calcium phosphate. When this happens the mitochondria stop or slow their ATP producing function which either kills or stuns the cell. At vulnerable points of injury there are a variety of protectors in different parts of the cell.

It is obvious that the needed minerals must be supplied in optimal amounts. Any toxic minerals or other poisons must be avoided or removed to allow for optimal functioning of the cell. There are over 5,000 known enzyme systems in the cell. New ones are being discovered all the time. The chemical reactions which represent the life of the cell must function according to the laws of chemistry and physics. Some of these chemical reactions are rapid and occur without help. Most of the conversions are too slow for the cells' needs. These reactions are catalyzed by special proteins designed for that purpose. Approximately 45% of these catalysts need a particular metal ion present in order for the reaction to occur at the proper speed. If the catalyst is not made in adequate amounts or if the wrong ion (poison) is on the catalyst the important chemical conversion will be too slow or will not take place. Two examples: Zinc is necessary in the breakdown of ethyl alcohol. Molybdenum is necessary in the conversion of sulfites to nontoxic sulfates.

The super energy exhibited by a normal two year old child is due to the super metabolic energy of his or her cells. The child has uncontrollable energy because all of his or her enzyme systems have redundancy and can operate at above needed speed (we wish we could package that energy and give it to older people). Old cells lose this enzymatic redundancy and are much more sensitive to mineral deficiencies or poisons.

Each cell has a predetermined assignment of work to do during life of the person. Different cells have different needs of nutrients and minerals. Diseases can hit one area of the body more than another, depending on deficiencies or excesses.

In summary:

1. Free radicals make holes in the cells. Poisons make the holes larger and more dangerous.
2. Magncsium, potassium and ATP from the mitochrondria run the pumps in the cells.
3. Calcium when in excess stops the pumps in the cells.
4. Holes in the cells will repair if poisons are removed and antioxidants supplemented.

Simplistically:

Our ship is slowly sinking.
There are three ways to help:
Plug the leaks!
Man the pumps!
Do both and quit worrying about it!

The Central Control of Body Repair

Stress can impose many and varied symptoms of illness on the body. Tranquilizers are often prescribed to patients when the number of symptoms exceeds one or two diagnoses.

The brain can be impaired just as any other organ including its job as central controller.

Stress may come from sources over which we have little control. So we must examine that over which we do have control. A healthy brain and a healthy body, physically can better deal with stress.

There are certainly psychological problems in life but we must first treat the brain as any other organ by de-stressing feeding and de-poisoning. This can reduce many symptoms the patient feels.

The brain is a fatty organ (you can call anyone a fat head and be right!). After you take out the water the remaining brain is over 60% fat. The fatty insulation of nerves and pathways in the brain direct the huge amount of electrical discharges that go on all of the time and hopefully prevent short circuiting of signals. Brain cells are very susceptible to lack of oxygen and poisonings of various kinds. The brain is protected by a special filtration system called the blood brain barrier. Without this system in place, simple food poisoning producing vomiting and diarrhea could be extremely dangerous.

The major energy source of the brain is sugar (glucose). Due to the large amount of unsaturated fats in the brain the anti-oxidant protection is very high.

For simplicity of understanding let's divide the brain into two sections:

1. The Thinking Brain (includes control over muscles and all the senses).

2. The Computer Brain, or the midbrain (controls systems which are usually considered out of our conscious command).

There have been numerous examples of people with severely damaged "thinking" brains which necessitated chronic life support. If the computer brain is not damaged then the person can survive in a vegetative state when taken off life supports.

Stephen Hawking in England has one of the greatest thinking brains of our time; his computer brain is badly damaged by Amytrophic Lateral Sclerosis (ALS).

We call the midbrain a computer since it really operates as such. If it were not a computer we would have accidents everywhere on our nation's interstate highway system. Joe Doaks stops at a rest stop buys and eats a candy bar. The subsequent rise in blood sugar is sensed by the computer which in turn signals release of appropriate amount of insulin from the pancreas. The insulin facilitates sugar utilization in the cells and may overshoot the need of this short blast of sugar intake. The blood sugar starts to drop rapidly. The computer notes the slope of the curve of the blood sugar drop which if left unchecked would cause severe hypoglycemia and possibly an auto accident. The computer midbrain sends a signal to a central receiving area called the hypothalamus, which in turn stimulates the pituitary gland, which in turn sends a signal to the adrenal gland to release adrenalin. Adrenalin (the lift you feel later) mobilizes stored sugar (glycogen) from the liver and avoids the disaster of severe low blood sugar. The blood vessels are also constricted by this adrenalin and the blood pressure is frequently elevated.

Our midbrain is a complicated sensitive chemical laboratory which monitors just about all of our bodily functions. The midbrain controls the hypothalamus which controls the anterior part of the pituitary. The pituitary controls cellular growth and development, thyroid gland, adrenal glands, sexual glands, etc. The posterior portion of the pituitary controls water loss through Antidiuretic Hormone (ADH).

The control that the pituitary has is governed by the hypothalamus in a manner that we call "negative feedback." If a woman reaches an age where her ovaries do not produce enough estrogen, the hypothalamus recognizes this lack of production and directs the pituitary to produce Folicular Stimulating Hormone (FSH), which in turn will try to stimulate the ovaries. When the ovaries do not respond adequately, the pituitary produces more and more FSH. When the FSH reaches a high level in the blood, it begins to interfere with bone remodeling, which can lead to osteoporosis.

Heavy metals or other poisons disturb the midbrain negative feedback control and can lead to lack of growth hormone, sexual hormones, thyroid, adrenal hormones and diminishing of the immune system.

The midbrain controls the alarm systems which are central to understanding stress. When we are threatened the midbrain directs the pituitary to activate the adrenal gland's sympathetic alarm system.

An example of sympathetic overload would be comedian/ actor Don Knotts doing his "fear and trembling" act. The body is poised to protect itself through adrenalin release, increasing blood sugar, increasing pulse and breathing. The opposite response would be a similar to a guru in a trance sitting on a mat. You place a firecracker behind him. Bang! He doesn't move. While this sympathetic response is normal and prepares you to defend yourself against injury, the defense process itself takes a toll, especially if the alarm mode is on for an extended time.

The alarm turns off appetite. You can't fight and eat at the same time. You are eating the finest meal that you have ever experienced and you receive a telephone call concerning the death of your closest friend. You cannot finish the meal. Stress turns down the immunity system in the same manner.

Various relaxing techniques can activate low sympathetic tone and mild parasympathetic increase which reveals the healing side of the alarm system. This is so necessary for whole body health.

Antiaging

Research on antiaging has increased greatly in the past twenty years. We all want to live longer and in good health. Basic research on certain hormone secretions reveals a gradual almost straight line decline with advancing age. I feel very strongly that the gradual decline of these hormone levels represents a gradual loss of what we talked about in "negative feedback" control from the midbrain computer. When we change lifestyle, nourish, de-stress the body, dc-poison and subsequently raise the immunity system, the sensitivity of the midbrain rises. The midbrain then sends the proper signals to the hypothalamus, which finally results in the more youthful levels of control hormones. The aging of the body slows considerably when these hormones are at a youthful level and in many ways the cells that are replaced will have more energy and by that definition will be younger.

After all of the healthful changes of lifestyle are made, if these hormonal levels are still below ideal, supplementation may be in order.

The hormones to be discussed in this section are cortisol, DHEA, growth hormone, melatonin, sex hormones, and thyroid.

The goal is to keep the antiaging hormones at a more youthful level. A seventy-five year old probably should shoot for somewhere in between 35 and 50 year age levels. It would seem that with the new information available now an individual age thirty or younger should be able, through careful yearly monitoring, to actually box that health curve and play tennis at 120+ years of age. The people starting later, if doing things right, should be able to reach age 100 in good health.

Benjamin Franklin was the first ambassador to Europe representing the new United States. When he returned he made the statement, "Eat breakfast like a king, lunch like a prince and dinner like a pauper." He had observed that people around him who were eating too much food in the evening were in poor health.

People taking their meal size in the opposite order had much better health. Was this observation important? We believe that this is extremely important and a major key to longevity. The argument goes as follows:

Everything we eat is foreign to our body with the exception of water and breast milk. We must break down food to its smallest components and then in our own liver reconstruct from these substances our own proteins, fats, etc. For argument's sake, if you swallow a small marble, it really isn't in you because you cannot break it down for absorption. The marble is just passing through the intestinal tract. A substance is in you only if it passes through the intestinal wall. People under stress (almost everybody) have "leaky gut syndrome" depending on stress level. Foods not broken down far enough remain in the body in a foreign state and stress the immunity system.

Stress is a killer but also part of life. It can be both good or bad depending how it is controlled and its effect on the body.

The Cortisol Curve

Cortisol is hydrocortisone produced by the adrenal glands (located above the kidneys). This is a major antiallergic, mineral-balancing and stress controlling hormone. It is secreted as our diurnal cycle hormone and delivered into the blood stream automatically on a fairly rigid time clock.

Figure #3 is an accurate cortisol curve for the person who rises at 7:00 A.M. and goes to bed at 11:00 P.M. Cortisol starts rising during sleep at about 4:00A.M. and peaks out at about 7-8:00 A.M. The cortisol level starts to crash at about 3:00 P.M. and completely crashes shortly after 4:00 P.M. The ideal times for food intake are during the times of maximum cortisol which are breakfast and lunch times. Then the body best tolerates food due to antiallergy control. We are recommending the largest daily meals during the period of highest cortisol protection. The very light suppers of soup and salad would be ideal. Snacks in the evening

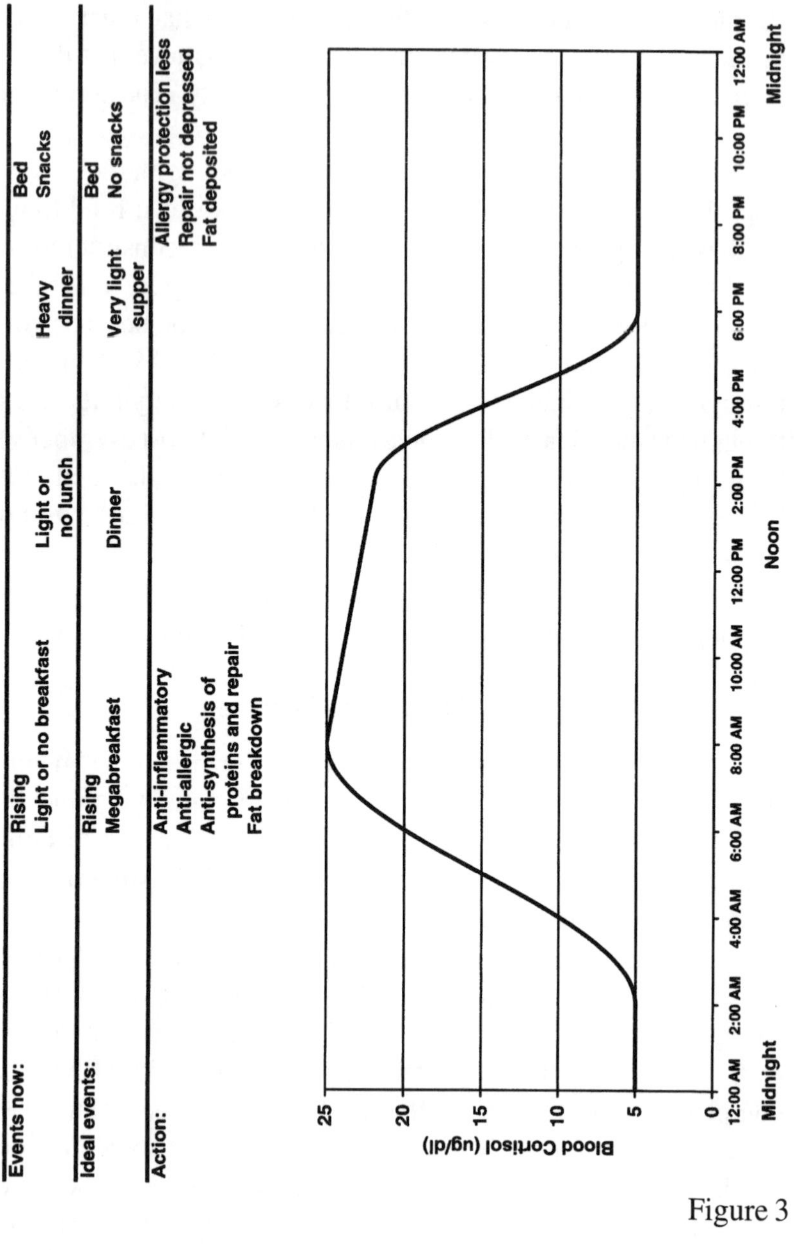

Events now:	Rising Light or no breakfast	Light or no lunch	Heavy dinner	Bed Snacks
Ideal events:	Rising Megabreakfast	Dinner	Very light supper	Bed No snacks
Action:	Anti-inflammatory Anti-allergic Anti-synthesis of proteins and repair Fat breakdown			Allergy protection less Repair not depressed Fat deposited

Blood Cortisol (ug/dl)

25
20
15
10
5
0

12:00 AM 2:00 AM 4:00 AM 6:00 AM 8:00 AM 10:00 AM 12:00 PM 2:00 PM 4:00 PM 6:00 PM 8:00 PM 10:00 PM 12:00 AM

Midnight Noon Midnight

Figure 3

should be avoided. The continued processing of food later in the evening delays growth hormone release during sleep. Food taken after the cortisol level drop tends to go into fat storage. (I would like to change this by extending the cortisol maximum level to 7:00 P.M., but unfortunately this is not going to happen). Only when we adjust our food intake to the system as it exists will we be amazed at how well it works.

The graph represents the time control relationship for a person who rises at 7:00 A.M. and retires at 11:00 P.M. The 12 hour cortisol level in blood is high approximately 4:00 A.M. to 4:00 P.M. and is low from 4:00 P.M. to 4:00 A.M.

The food taken during the high cortisol period is much better tolerated and the potential for allergic stress is much lower at that time.

When we change the diet timing from breakfast into a mega breakfast, lunch into a dinner, and dinner into a very light supper, the body handles food in a much different manner.

The food taken after the cortisol level has fallen (4:00 P.M.) tends to go into fat storage and is a major stressor on the body.

We realize that you will "go out" periodically for a larger evening meal and that probably is okay, but routinely we should have only a very light supper such as soup and salad. Meat is the most complex protein to be broken down and should be reduced for the evening meal.

Why do we sleep? This is the body's time to do most of the necessary repair work. The release of the growth hormone overnight tells the cells to divide, replacing dead cells. Sleep is important. After attaining your full height, growth hormone becomes the cell replacement hormone.

DHEA

This adrenal steroid is a most remarkable prolongevity hormone. *Dehydroepiandrosterone* is this hormone's official name and you can understand why we call it DHEA. DHEA is taken orally and therefore very desirable as a replacement.

The hormone can be a precursor to just about any of the hormones the body can use. It has been thought of as only a precursor to the sexual hormones but its actions are beyond that. When you are young the amount of DHEA is huge and at about 80 years of age it is down to about 10%. You have heard the term "he is 80 going on 50" or "he is 50 going on 80." Health age and chronological age are not the same and both Growth Hormone and DHEA are good markers of true health age.

DHEA clearly supports the immunity system. Low DHEA levels are correlated with degenerative diseases of all types. Chronic fatigue and even obesity may be helped by continuous supplementation.

A test for DHEA-S found in the blood correlates most clearly with the true level of the hormone in the body. This hormone is basically nontoxic; however, the blood levels should be monitored and supplementation given according to the need. As in all hormones more is not necessarily better.

Human Growth Hormone

"My, how you've grown!" How many times has that been said of a child you haven't seen in a 6-month to year period. Yes, Growth Hormone (HGH) did in fact come into the body of that 12-year-old in huge spurts during that time period.

We take growth for granted and once we get our full height we assume that growth is complete. The phrase "youth is wasted on the young" probably is somewhat true. During that growth period, special attributes can most easily be developed. The time of rapid growth is the most opportune time for skill development in learning languages or special sports abilities.

After we reach our full height we really should rename HGH the "Cell Replacement Hormone" because that is its function.

HGH comes from the pituitary under the control of the hypothalamus. While many animal-derived hormones are used satisfactorily in standard medicine, HGH is remarkably specific and only HGH is effective in the function it has in human cell division. Bovine Growth Hormone used in the beef industry is economically profitable in beef production; however, it has no known growth effect on humans.

A congenital deficiency in HGH produces the rare condition called Progeria, when a child of 12-14 dies of old age. The child has all of the characteristics and diseases usually associated with extreme old age.

If lack of HGH can produce Progeria many believe that HGH supplementation could be the "fountain of youth." HGH is produced by the pituitary in spurts primarily at night. In order to test HGH directly it would be best to take blood samples all during the night. HGH goes to the liver and is converted to Somatomedin-C (which is also named *Insulin-like Growth Factor-1*). Somatomedin-C is a much more stable compound and reflects the level of HGH the body maintains.

Cells, depending on the type in our body, die on a routine schedule. 90% of all cells in our body turn over to new cells every three years. Certain brain, nerve and heart cells do not turn over. Red blood cells live about 120 days while white blood cells may live only hours to a few days. Skin and intestinal lining cells turn over rapidly. During our growth period the high level of HGH ensures that there is more cell replacement than death of cells. After we have grown to our full height the cells replaced are supposed to be equal to the cells dying. In people who do not obtain their optimum health, the decline in HGH from 35 years of age on is almost a straight line to age 100.

There are many variables and the following formula is not a guide to health; however, it is useful in beginning to understand HGH:

$$\text{Cellular Replacement Age} = 115 - \frac{\text{SMD-C}}{4}$$

This is a guide from age 35-100 only. Blood Somatomedin-C is in NG/ML.

Most people reading this manual have heard about the attributes of HGH. Interest in this hormone is recent since the Dr. Daniel Rudman study of 1990 reported in the New England Journal of Medicine, "Effects of Human Growth Hormone in Men over 60 Years Old."

This study of only six months in length showed an increase in skin thickness, muscle increase and fat reduction plus a sense of increased well being in the male subjects.

This short study didn't reveal the overall cellular repair seen with longer use. Just about all of the signs we associate with aging were affected for the better when SMD-C is improved. The aging process was stopped and partially reversed.

HGH replacement is by injection. This hormone is given using a very small gauge needle under the skin preferably four to six times per week.

Does everyone need HGH replacement? Absolutely not! Everyone needs to maintain the HGH at an optimal level for their age. When a person has proper nutrition and has had the poisons removed from the body, the HGH will usually normalize. If it doesn't normalize then perhaps HGH replacement is in order.

Melatonin

Interest in melatonin as a miracle hormone that will solve all problems has grown in recent years .

We are a very complicated animal whose lifestyle does not lend itself to being held captive in cages and given a single substance to see what happens. Studies on mice have shown an increase in life length of about 25% in their life span with the use of melatonin.

Is this going to happen with us? I doubt it; however, it is clearly something that is a factor in the life extension puzzle. Melatonin is an exceedingly strong antioxidant that is released from the pineal gland located in the middle of the brain. This gland's secretion of melatonin has been known for many years but its profound functioning gradually reduces with aging.

Until recent years the greatest value of the pineal gland was finding whether it was moved off of its central position by a possible brain tumor. Radiologists love the calcified gland.

There is an important connection between adequate melatonin and an enhanced immunity. Melatonin has been shown to be able to regenerate the thymus gland. The body is being challenged continuously from outside by invaders such as viruses, bacteria, fungi and parasites. In addition we are being challenged internally by malignantly changed cells. If that is all that melatonin did, that alone would be enough to make it a vital part of our longevity goal. It does more.

The pineal gland is part of our internal clock which controls our circadian cycle. The pineal gland influences the cortisol cycle since the use of melatonin is helpful in correcting jet lag. The pineal gland normally is sensitive to the light and dark and secretes the melatonin during the early hours of sleep. As we age it is more and more difficult to have adequate good sleep, because the production of melatonin decreases. Insomnia is an increasing problem "solved" by an increased use of melatonin cycle control.

Sexual Hormones in General

While the sexual hormones have a lot to do with the sexual differences between men and women and their ability to reproduce, the story is more complicated and definitely involves what we are looking for in "boxing the health curve."

The pituitary is controlled by the hypothalamus which delivers a regulator called FSH (Follicular Stimulating Hormone). This hormone stimulates estrogen from the ovaries or testosterone from the testes. The adrenals produce both testosterone and estrogen in both men and women. The testosterone level in the women, although only about 10% of the males' level, is important. The estrogen production in males is equally important for the men since estrogen is needed for brain function.

The Estrogen Story

As the ovaries mature in young girls there is finally enough estrogen produced to start the secondary sexual characteristics of the woman. These of course are breast enlargement, fat distribution on the hips, pubic hair, etc. The estrogen starts the growth of the lining of the uterus (endometrium) which grows to a level sufficient for implantation. Progesterone comes into play secreted by the ovaries which softens the lining and facilitates the possible implantation of a fertilized egg. If a fertilized egg implants in the lining, the ovaries produce a larger amount of progesterone to help maintain the pregnancy. If there is no implantation the ovary stops the progesterone secretion and the lining is shed, creating a menstrual cycle. If the woman is having regular menses in both timing and normal amount we generally consider that her hormones are satisfactory. Birth control pills reduce the amount of blood loss and that reduction increases the incidence of vascular problems.

When the woman stops menses at menopause she starts retaining the iron which increases her risk for free radical damage. The hormonal reduction in estrogen from the ovaries may produce vascular instability called "hot flashes."

Symptoms of menopause vary considerably but can be very mild to severe. Symptoms include depression, vaginal dryness and atrophy, memory loss, fatigue, mood swings, breast atrophy, insomnia, loss of libido and others.

These symptoms and conditions suddenly are upon the woman. The next question is what to do about it, if she is convinced that estrogens are associated with an increased cancer risk.

It is important to realize that while there are studies showing an increase in uterine cancer with estrogen usage, that risk is well controlled when progesterone is given at the same time. If estrogen replacement is made in the form of triest (80% estriol; 10% estrone and 10% estradiol) the cancer risk is reduced further. Estriol, while being a weak estrogen, has actually been shown to have anti cancer effects. If the triest is given daily and the progesterone is also used at the same time the endometrial lining doesn't develop and therefore there is no period. If it is cycled, a "phony" period may well develop. This technique was probably devised by male doctors. Most women who develop a spontaneous re-starting of menses (say one year after menopause) tell me "I'm glad I'm getting younger through life style changes but I wanted to keep that area a little older." It is interesting that at menopause women join the ranks of males in retaining iron. They have been remarkably protected from heart disease until menopause. By following the iron storage (ferritin) buildup and donating blood periodically the increased heart disease risk can be lessened.

The Testosterone Story

As you have noted by now all of the hormones go down with advancing age and are considered a natural part of life leading to inevitable death. While the sudden loss of menses in the woman is certainly a protection against later age pregnancy, the male has no obvious similar awakening to aging. As long as he has morning erections and is able to perform the sexual act he usually tries to forget his sexual aging.

Testosterone has bad press. Testosterone replacement therapy does increase muscle mass because it is an anabolic steroid. While both estrogens and testosterone are prescription controlled, only testosterone has DEA restrictions. The potential for misuse is huge. There have been a number of self-medicated tragedies which have discouraged people regarding its use. When we talk of hormone replacement we are talking about returning the testosterone to a level of what the man and woman had at about 35 years of age. We are not recommending a huge dose for some hidden agenda of excelling in sports.

The DEA visited my office and told me that I could have my lawyer with me if I wanted. I asked why I would need a lawyer and they said "because you use a lot of testosterone by injection." I agreed that we used a lot of testosterone as replacement therapy because a lot of men and women are very low in testosterone. After bringing patients' records showing the low blood levels and then later ideal levels of testosterone in the patients, the DEA agents said, "You are the only doctor using it properly." The DEA gave me a certificate of inspection and thanked me. (I find it difficult to believe what they said; however, it does point up the fear of testosterone replacement).

I find nothing dangerous in returning testosterone or other hormones to a younger adult level after life style improvement has been made.

Testosterone therapy in isolation may not do all of the things mentioned but it does have a definite effect on the ability to have sexual intercourse and on muscular strength. Fatigue is reduced, depression lifts, mental sharpness increases. A reduced rate of osteoporosis, and other benefits were shown from the proper amounts of testosterone. Studies with testosterone also showed reduction of the risk factor fibrinogen, lowering insulin resistance, improving libido, as well as diastolic blood pressure lowering.

There is no evidence that testosterone produces cancer of the prostate. The prostate produces a specific protein called Prostate Specific Antigen PSA. This is normal at less than 4 ng/ml in blood. Over 4 this may be just BPH (Benign Prostatic Hypertrophy); however, over 10 there is a possibility of cancer of the prostate. A digital prostate exam can usually evaluate the prostate with higher PSA. Prostate cancer may well be sensitive to testosterone therapy and therefore the prostate should be monitored at least yearly in males on testosterone therapy.

While we see a lot of patients with prostate cancer coming in for consultations, the development of prostate cancer in patients already on a preventive lifestyle is very rare.

Review of the Thyroid

The thyroid gland is in the front part of neck just below the "Adam's apple", in front of, and extending beyond both sides of the trachea. On physical examination a normal thyroid can barely be felt.

This gland produces thyroid hormone. It is controlled by the hypothalamus which in turn controls the pituitary. The pituitary sends out thyroid stimulating hormone (TSH), which instructs the thyroid to produce more thyroid hormone.

The cells in our bodies have receptors on the outer membranes of the cell. These receptors are designed only to be activated by the specific hormone configurations. This hormone is a "key" to the lock. The function of the thyroid hormones is to increase metabolism. Metabolism is dependent on energy production and this "key" is the signal for more ATP to be produced in the mitochondria.

The body temperature depends somewhat on the person but the number 98.6F (37C) is considered the upper limits of normal. Most people are cooler than this temperature but not by a lot. Temperature below 97.6, possible 97.8 F on an average is probably less thyroid production that one needs.

The hormone T4 goes from the thyroid gland to the liver where it is converted to another thyroid hormone called T3 and an optical isomer which is called reverse T3 or RT3. T3 is the real key to turning on the metabolism of the cell while RT3 just blocks the thyroid receptor without turning on metabolism.

Thyroid supplementation is definitely different than the other Hormone Replacement Therapies. The thyroid production does not decrease directly with age despite the fact that hypothyroidism is extremely common. We have found that heavy metal poisoning disturbs the hypothalamic pituitary connection and when this is corrected the thyroid function improves and the RT3 blockage is generally reduced. It is interesting that when metabolism is down (hypothyroidism), that wonderful building block, cholesterol, goes up. Millions of people are being treated for cholesterol elevation when the problem is really low thyroid function.

The classic symptoms of low thyroid are fatigue, depression, weight gain, fluid retention, lower body temperature and intolerance to cold. The classic symptoms of high thyroid are just the opposite: nervous energy, sense of disaster, weight loss, higher temperature, rapid pulse and intolerance to heat. While severe hyperthyroidism can kill, the usual brand of hypothyroidism only makes people miserable until it is corrected.

Since we have said that normal cell energy is our goal, excess thyroid is not the ideal way for weight loss.

Parathyroid Glands

The Parathyroid Glands are very small glands residing behind the thyroid gland. There are four or more of these glands and they secrete parathormone which regulates calcium and phosphorus metabolism.

The average person with a diet of excess in phosphate creates a condition of dietary hyperparathyroidism which is a major influence on removing calcium from bone and allowing it to go into soft tissue.

Excess phosphorus comes from diets high in meat and soft drinks. We do not need a large intake of calcium as long as the phosphorus is not in excess. The classic Japanese diet which contained only 450 mgm of calcium daily did not produce osteoporosis because the diet only contained 600 mgm of phosphorus.

Brown Fat Gland

(The gland that you probably didn't know existed)

When a bear hibernates, the fat stored up in the summer is gradually converted to water and heat. Muscular activity is not needed in this conversion due to the Brown Adipose Tissue gland (BAT). When we get cold we generally start to shiver which is muscular activity used to generate heat. In humans the BAT is located between the scapulas on our back and may also be located in small amounts in other locations. People who live in cold areas have more BAT than people who live in a mild climate. BAT converts fat to heat and water without muscular activity.

Very small amounts of active BAT could increase the fat breakdown rate by 20-25% which would make a huge difference between maintaining weight and gaining 40-50 pounds per year. Adequate thyroid activity is necessary for BAT to operate properly.

BAT is increased by sympathetic stimulation (cold and exercise work well to stimulate BAT). Melatonin acts directly to stimulate BAT. Capsicum (cayenne pepper) and L carnitine also have been shown to stimulate BAT.

We have discussed previously about the gradual decrease in hormonal activity as we age. The adult onset diabetes, gradual increase in our weight, and decreased exercise activity are all factors in making BAT influence reduce as we get older. I believe that the increasing load of heavy metals and other poisons in our aging body also continues to lower the BAT. Medical research on understanding the relationship of HGH, melatonin, thyroid, sexual hormones, adrenal hormones and BAT may reveal the true nature of obesity. It is <u>not</u> simply eating too much.

The Gastrointestinal Tract

It is amazing how well this system works considering how it is insulted on a day to day basis in the lives of the average owner.

Food is chewed. Saliva mixed with the food starts breaking down the starches. The food is swallowed and enters the stomach through the cardia. This opening into the stomach can go into a spasm called "cardiospasm." This can be very painful in the mid-chest and frequently is severe enough to frighten people into the hospital's emergency room, thinking it is a heart attack. The stomach now gives the food an acid bath. This starts the major breakdown of the food and basically sterilizes it by killing bacteria. When there is stomach discomfort or indigestion, this may well be a lack of stomach acid. The term "indigestion" seems to always be "acid indigestion." While a popular practice, giving antacids is not necessarily the proper treatment. The food then moves to a bile and pancreatic enzyme bath in the beginning of the small intestine. The bile emulsifies the fats which are then broken down with the pancreatic enzymes. The enzymes are alkaline and they neutralize the acid from the stomach. The food now has a ph of 7-7.5 and starts to break down further, releasing the food as broken down components which the body recognizes. Bacteria of Lactobacillus acidophilus group help in the breakdown of food. B vitamins are released. Bacteria keep the intestinal wall clean and keep "bad" bacteria down since they are L. acidophilus (acid loving) and most pathogenic (bad) bacteria do not like acid. The food then moves to the large intestine which processes food further. Bifidus, as good bacteria, stimulate fluid release from the walls of the large intestine.

Survival Skills

When you are in a threatening situation with lack of water on a dangerous crossing of the desert you could obviously die of dehydration. The midbrain sends signals to the posterior part of our pituitary gland which releases Antidiuretic Hormone (ADH). ADH reduces urination to a minimum which conserves water. The large intestine extracts water from our feces. Perspiration is lowered but maintains the cooling effect. If you don't drink enough water the body is always secreting excess ADH and this is undesirable.

You would like to lose weight and you decide to eat less food. Because survival skills were designed to function back in the time of early man when food was not readily available, the metabolism rate controlled by the thyroid drops. By burning up less food weight loss is much more difficult. When you eat less, deficiencies in vital weight reducing nutrients are more likely. If you are in a situation of reduced food intake and you don't want to lose weight, then obviously this survival skill is needed.

Women live longer than men by 5 1/2 years due to 30+ years of menstruation. The more iron early mankind stored the more chances of living longer in case of a major injury and severe bleeding. The life expectancy was low anyway so it was definitely a survival skill. Iron (ferritin stores) is one of the major catalysts producing free radical damage and aging.

Dr. Jerome Sullivan showed that the accumulation of ferritin in men was the sex difference in the prevalence of heart and vascular disease. High ferritin was reconfirmed as a major risk factor in a study from Finland in 1992.

There are lots of examples of survival skills being a problem and we want to start understanding how the system works so we can take advantage of our knowledge if we intend to reach 120 years of age in good health.

THE TEN BASIC PRACTICES FOR GOOD HEALTH

1. Increase healing rest and reduce mental stress.
2. Exercise the body properly.
3. Eat the correct quantity and variety of foods.
4. East less food in the evening.
5. Reduce or eliminate environmental poisons, including smoking and excessive alcohol.
6. Drink more water free of poisons.
7. Eat organically grown, non-poisoned food.
8. Avoid unneeded drugs and surgery.
9. Read labels on everything that goes into your mouth.
10. Take adequate antioxidant supplements.

II. MAINTENANCE

Introduction

In the first section we discussed how the system works, so that you can begin to understand the importance of doing everything possible to maintain and strengthen the energy of the cells. Proper maintenance prevents disease. Lack of proper maintenance leads to the need for troubleshooting (in the third section of this manual). If you never change the motor oil in your car doesn't this produce a problem? We are a long lived animal and the statement "If I had known I would live this long I would have taken better care of myself" is true. Poisons are also discussed in this section. We are able to "tolerate" a certain amount of continued poisoning unknowingly.

Depending on our genetic makeup it is amazing how "well" some people can appear on the outside, yet really have serious correctable biochemical and bioelectric problems. Obvious external problems arrive late in life which may be almost irreversible (see Lemon Law in Troubleshooting).

In this section we discuss water, fuel, air, stress, poisoning, testing, supplementation, and practical tips in maintenance and economics.

How do you do all of this on a continuing basis? We are all "creatures of habit." Look carefully at the recommended practices which deviate the most from your life style. Concentrate on those changes first. True health will improve over a reasonable period of time. Do not be overwhelmed by trying to change everything overnight; it's not possible. Patience. Change gradually.

Everyone agrees "you are what you eat." No one asks you what you eat, therefore, no one knows what you are! By this principle everyone is unique, since no one has the same array of poisons fouling up the self-repairing systems. The supplementations recommended are for the average person. It is best to have the supplementation supervised by a preventive-oriented physician who has done adequate preventive-oriented biochemical testing.

Water

Drink enough pure water. Water is the most abundant substance in our body and the most neglected. We can survive for possibly thirty days or longer without food, but only about three to five days without water.

Water is basically the transporter of all nutrients throughout the body. Water is involved with virtually all the body's biochemical reactions and vital for excretion. The average body with a lean body weight of 50-70% will probably contain 40-50 quarts of water. Water is critical for body temperature control, and depending on activity and humidity, the loss of fluid can be much greater than anyone realizes until a problem develops. The sensation of thirst is very finely tuned in some people and not in others (dehydration is one of the most frequent conditions in the hospitalized elderly). For people of all ages water should be drunk throughout the day and not just when one is thirsty.

Satisfying thirst with coffee, tea and colas (all of which may have a diuretic effect), does not correct the body's water shortage. These beverages are not pure fresh water.

Drink most of your water between meals. Too much water taken with meals dilutes stomach and pancreatic enzymes and impairs the processing of food.

If you don't like the taste of pure water, use a lemon or lime slice in the glass or use herbal tea.

While clean water through the tap has been the most important health improvement in the world, we still can do better than drinking city water. Contaminations from many sources and the pipe delivery system in a community make the chlorination of the water probably mandatory. Chlorine has been linked to many health problems, including cancer and heart disease, so keep tap water to a minimum. Chlorine is ineffective against Giardia and Cryptosporidium parvum which periodically are reported in large urban water supplies.

Water from wells can be wonderful or terrible depending on the water table source. Know your source and if you don't, test the water before accepting this as your daily source of water.

Distilled water is water that is recondensed from steam and flat tasting but pure. We will not receive our mineral requirements from this water source, but you may want to use distilled water for brushing teeth or making coffee, tea, or cooking. Distillation may not take out some volatile organic compounds sometimes found in ground water.

Most of the name brands of bottled water are very carefully monitored as far as any toxic substances are concerned. Read labels on the bottles and ask for analysis from the company. Glass containers are preferred, especially for persons with chemical sensitivities.

Water filters can be activated carbon filters which remove odors, color, taste, and contaminants such as organic compounds, mercury, and chlorine. Filter replacement frequency determines your overall cost. Filters can become a breeding ground for bacteria, etc., if not replaced frequently.

Pitcher-type filters may remove lead, copper, chlorine and sediment but do not take out asbestos, nitrates, sulfates, viruses, Giardia, or Cryptosporidium. They can be inexpensive, but the filters need weekly changes.

Reverse osmosis water filters are probably the most efficient remover of undesirables in water. The semi-permeable membrane system needs professional installation and periodic maintenance.

Ultraviolet light destroys pathogens and viruses but does not affect other undesirable water contents. As a sole source of water purification it is inadequate.

Water softening systems remove high calcium and other minerals in the water and substitute sodium for these minerals. These systems solve dishwasher and laundry problems, but they are <u>not</u> drinking water. A water softening system should be connected only to dishwasher and washing machine water sources.

Fuel—What Do You Eat?

When we talk about good food we don't always know what is good food. Pesticides, poison and processing have stripped foods of value. Hormones and antibiotics in meats are damaging to us. Only nitrogen, phosphorus and potash are returned to the soil, disregarding ultimate mineral needs of consumers of food.

Medical school education made no emphasis on diet; hence, dietary history is seldom part of a patient's medical history. Crisis-oriented medicine does not consider a patient's diet as germane to a problem unless food poisoning is suspect.

Keep your own dietary history of everything you put in your mouth for two weeks. If you are what you eat--get to know yourself!

We are an omnivorous animal, which means we can eat anything. If you look at our teeth you will find that they are that of the plant eating herbivores with four token canine teeth. We probably can say from that observation that we are basically vegetarian by nature. Even today a large number of people are doing what early man did, and that is hunting and gathering. The emphasis should be on gathering of vegetables, fruits and eggs, nuts, grains, etc. Today the majority of the world's population shops or gathers food on a daily basis because of lack of refrigeration.

Unless the food is poisoned or low octane (purified carbohydrates such as sugar, white flour, and alcohol) there are no obvious short term effects on health, hence standard medicine's limited interest. Erroneously, many people do not believe there is a relationship between food and disease!

Although I'm not going to try it, I could probably drive my car on 90% gasoline and 10% kerosene for a couple of years before serious engine problems occurred. Likewise, you can, and many do, run the body on a 'junk food' diet for years before a disease process is obvious.

Carbohydrates

Complex carbohydrates are basically vegetables and fruits and should make up the bulk of everyone's diet. Grow or buy organically grown food! We are being exposed to ever-increasing amounts of pesticides in our food. Make every effort to keep these foods to a minimum. Fruit in its original form is best since the purification (fruit juice) reduces fiber and the rapid adsorption of even complex sugars can be a problem.

You should try to include a variety of fruit in your diet since we want to take advantage of the variety of solid minerals which this affords. Tropical fruits such as ripe mangoes, pineapple, kiwi fruit etc., should be added to our fruit list.

Grain is the seed-bearing fruit of grasses. Whole grains contain complex carbohydrates, protein, vitamins, fiber and phytochemicals which are all necessary for good health. Processing removes many of these nutrients. When flour is enriched, that usually means just about everything was removed that was important. Some grains of importance are amaranth, barley, buckwheat, bulgar, kamut, millet, oatmeal, quinoa, rice, spelt, teff, wheat, and wild rice.

Vegetables, organically grown if possible, should be the main bulk of your meals. We recommend that you eat twice as many vegetables as fruit.

Vegetable protein sources come basically from beans, peas, lentils, and soy products. Carotene rich fruits and vegetables are dark green and deep yellow, including carrots, apricots, yellow peaches, sweet potatoes, winter squash, pumpkin and others. Cruciferous vegetables include cabbage, broccoli, cauliflower, brussels sprouts, kale, chard, mustard greens, and kohlrabi. In just about every vegetable known there are important phytochemicals that can be used to the body's advantage. One of the most important items in diet is garlic.

A huge amount of good scientific research has been done on the immunity enhancing ability of garlic which increases T lymphocyte activity, macrophage actions, interleukin-1 and natural killer cells. Garlic is our major sulfur source from the vegetarian side. Without adequate sulfur in the body, heavy metals are not able to be excreted well. Since mercury poisons the thyroid control and low thyroid hormone level increases blood cholesterol levels, perhaps this is the reason that garlic has been shown to decrease cholesterol levels by about 9 to 12%.

Sugar, white flour, soft drinks, alcohol are all purified, or refined, and we consider them "low octane fuel." Low octane fuel makes the engine knock due to too rapid an explosion in the cylinders. While the complex foods give sustained energy, the "knocking" of low octane fuel is producing a whole generation of ill children as well as adults.

When children in a sugar-growing area grab a sugar cane stalk and chew it, the sweet juice which contains vitamins, fiber, and other nutrients is not a problem. They are chewing on a few inches of crude sugar cane. A candy bar represents perhaps a yard of sugar cane sugar, from which all of the crude and important nutrients have been removed. The behavioral changes and decreased learning ability of just about anyone, particularly children, should be obvious to any parent or adult.

The annual per capita intake of sugar in the United States is over 150 pounds per year (63 pounds 100 years ago). This translates to over 30 teaspoonsful of sugar per day. These empty calories need B complex vitamins and minerals to metabolize.

Sugar is added to prepared foods so the consumer's taste buds will be satisfied and the product sells.

White flour is in just about everything if you don't watch out. A doughnut is a great example of white flour and sugar put together as a non nutrient. White bread is usually worthless. <u>Read labels</u> of contents on cans, on packaging, on bottles of <u>everything</u> you use. The first item listed is the highest amount of ingredient; the last item is the least amount of ingredient.

When you carbonate a beverage you add CO 2 and water, producing carbonic acid (H2CO3). This doesn't really taste great so the manufacturers put in a phosphate buffer into the drink and now you just taste the pleasant feel of the bubbles. The phosphorus in the soft drink stresses the calcium and phosphorus balance in our body which is controlled by the parathyroid glands. This leads to dietary hyperparathyroidism which is a major factor in osteoporosis and causes increased chances for vascular disease.

The sugar in the pop is bad enough but aspartame is even worse. The breakdown products of aspartame are formic acid (bees and ants sting you with it), formaldehyde (poisonous pickling agent), and wood alcohol (can produce blindness). We don't need this.

If you really "need" a sweetener for your tea or coffee, you might try stevia. This herbal contains natural compounds called stevioside and rebaudionides that are 150 to 400 times more sweet than sugar. This is so sweet that only 3-5 drops in a cup are needed. It can be purchased at a health food store.

Proteins

Meat is a good source of protein for you! Look closely at this.

We need proteins in our diet and I don't believe that you can get all you need entirely through a vegetarian diet. Our meat supply has gone downhill rapidly in the past twenty years and most available meats are suspect. This manual is too general to rediscuss the problems with our meat industry or a pure vegetarian diet. Read Diet For a Small Planet and Fast Food Nation. When meat is decompartmentalized (a fancy word for ground meat), the value of the proteins goes down as oxidation of the meat is just about complete. Meat should be intact in the cooking process and not overcooked. Excess cooking produces heterocyclic amines (HCAs). These compounds are found in the crisp black crust on well done meats, fish or poultry. A 1998 study sponsored by the

National Cancer Institute showed that women who eat overcooked meat with a blackened crust have five times the risk of breast cancer compared to women who eat medium or less cooked meat.

If you are able to obtain organically grown meat, the effort is worth it. We need to take less meat in general and eat it primarily at noontime (see *timing of meals* under "Stress").

The cholesterol theory of vascular disease is a total myth, but there are some interesting rules regarding eggs. We should use "free range eggs" coming from chickens that are not in concentration camps. The yolk should be intact in the cooking process. The egg can be poached, soft boiled or fried over easy. The white of the egg contains all of the iron, and when you scramble or make omelettes the yolk becomes oxidized during cooking and we lose our nutritional value of the eggs. The eggs can be hard boiled, and while the yolk is not oxidized the proteins do lose some of their value. You can and probably should eat two or more of these free range eggs per day. Eggs are our major meat sulfur source. Ever smell a rotten egg?

Fish is a major protein source and certainly can be an important food item. Where does the fish come from and does it contain heavy metals such as mercury and arsenic? Farm raised fish can be great only if the ground water at the farm and the fish food is good. Fish taken from rivers downstream from major cities should be considered toxic until proven otherwise. If you have a reliable and consistent source of fish, the flesh of the fish can be tested for toxins (see laboratories listed in Appendix).

Swordfish, tuna steaks, shark, and some salmon tend to have more mercury than others. Scrod (halibut, flounder, cod and haddock) from the North Atlantic usually are satisfactory.

Fats

Fats are bad, right? Wrong! Fats are vital to our life. The only problem is the quantity and quality of fats taken in.

The average American has 45-50% of his/ her diet in fat and it basically comes from the hind end of a poisoned cow.

All of our cell membranes are made out of fatty acids. Our brain and nervous system insulate with fatty acids and our regulatory hormones are made from fats.

We need to keep the fat intake at about 25-30%. The fats should be good fats that can be used by the body. Butter is good in limited amounts (linoleic acid in butter is an essential fat). All margarines should be avoided since they are made from hydrogenated vegetable oils. Hydrogenation produces a fat which doesn't occur in nature and cannot be utilized in the body.

Beware of foods labeled fat free. Vegetable oils when hydrogenated become fats which are not compatible with the body. However, because these oils do not come from animals and oils do not contain cholesterol the FDA apparently allows the manufacturers to label the products "fat free." These products are not "fat free."

When we test patients we check for essential fatty acids (EFAs) and try to obtain the ideal balance between Omega 6 and Omega 3 fatty acids. Omega 6 EFAs come from leafy vegetables, legumes, raw nuts, seeds and safflower oil. Omega 3 EFAs come from cold water ocean fish, soy, walnuts, and flax seed.

Canola and corn oils provide both, so mix a cold pressed vegetable oil (50% oil-50% butter) with your butter and some vitamin E to retard rancidity. It's a more healthful spread.

We need a moderate amount of good fat. Low fat diet programs such as Pritikin and Ornish are to be avoided for any prolonged use. If fats were really terrible we would see serious disease in the Eskimos, who eat a huge percentage of their diet in blubber. The Eskimos have very little heart disease and are almost completely protected unless they eat processed food and smoke cigarettes.

Salt

The addition of salt to our diet has been greatly reduced over the past few years because we've been told that salt is bad for us. But the body needs salt in the right amount. Excessive salt is easily excreted as long as the body has enough water. The iodine level in food is low in areas away from the ocean coasts. The major source of iodine for many people is through iodized salt.

When exercising, not only do we excrete water but also many minerals, the largest amount being salt. Since perspiration is the major heat regulatory system in the body we do not want to be low in salt when we exercise or when the weather is hot.

Most of our salt intake (actually about 75%) comes from processed foods. Check labels. Salt makes food taste better and is usually added in processed foods. Try to control your salt intake yourself. There is no question that excess salt can cause blood pressure to rise in some people. If you have difficulty controlling salt levels then be sure to include this in testing.

Salt and water are so critical for health that we have no generic advice in reference to salt except that the proper use of antioxidants is the key to whether salt becomes a serious problem.

Sodium is an extracellular (outside the cell) mineral and becomes very toxic to the cell when it is inside the cell in excess amounts.

The food additive monosodium glutamate (MSG) is a transporter of sodium intracellularly and therefore is toxic. This "flavor enhancer" is the cause of the so-called "Chinese Restaurant Syndrome." If a person already has excess intracellular sodium, MSG is occasionally a killer. Many food processors hide the presence of MSG by using the term "hydrolyzed vegetable protein" or "natural flavorings." Some of the symptoms of MSG reactions are burning sensations, tightness in the chest, headaches, flushing, pain, asthma, joint pains, dizziness and insomnia.

Miso, the fermented soybean food additive, is a central player in the "Macrobiotic Diet." It contains sodium and other minerals the body needs to keep sodium from becoming a problem. Use it in cooking and in soups since it is one of the best solutions to salt's "double-edged sword."

Air

How is the air we are breathing? The world's supply of oxygen is going down due to the rapid reduction of the rain forests and trees in general. We have to counter by keeping the air we breathe as clean as possible. Besides the huge list of potential chemicals that we can see in the EPA's poison lists we have to protect ourselves from particulates from all sources. This can be difficult, especially in an urban environment. The rural dweller has to be careful from air-borne insecticides in an otherwise less toxic environment.

If our immune system is good we can survive the onslaught of toxins; however, it isn't going to be easy. When the immunity system is poisoned we become more and more like the canary who was used as an early warning for bad air in the mine.

Humidity in the home is a major consideration of respiratory comfort. During winter in colder climates a humidifier is essential for health. Air conditioning during the summer may be effective for many only if coupled with an efficient electrostatic type air cleaner. Many homes are toxic due to mold, dust mites, and disintegrating insulating materials. We have found that since the body uses a similar antiallergic adrenal steroid defense system both for respiratory and food allergies, de-stressing the body with the removal of food allergies can greatly help respiratory impaired patients.

Stress Control

There are thousands of books on relaxation and stress control. Some methods work on one person and will not work on others. There are many disciplines in the Orient for stress control. Yoga, Transcendental Meditation, Jin Shin Jyutsu, etc., all are very helpful in turning off our internal alarm system. Seek out those which fit you and which you will follow regularly.

It is shocking to me to find how many so-called severe psychiatric illnesses are turned around completely through de-poisoning and proper diet. Some mental stress may be easily reduced by something simple such as a hot and cold alternating shower. Perhaps expert massage therapy will help. Stress doesn't go away unless you are willing to let it go away. Family, work, financial problems are not solved by worry. If worry solved problems I would suggest that you worry. I know of no chronic use of a drug that will solve mental stress.

The immunity system is seriously impaired by the extra adrenal compounds released due to chronic stress and we feel that this is a major factor in the development of immune suppressed diseases such as cancer. We find that almost everyone with cancer has a history of chronic hatred toward someone. This could be a family member, boss, someone suing you, etc. This continuing stress is destructive and must be stopped for your own protection. Anger and hate are self-destructive.

One cannot process food under mental stress. Malnutrition from lack of processing of food is a common problem of people under chronic stress. Frequently digestive enzymes are needed for these patients. There is a test for adequacy of these enzymes.

Exercise

Recommendations for exercise should be made on an individual basis. A generic recommendation for most people as a goal to work up to would be a half hour fast walk every other day. One does not have to do strenuous exercise to be in good health. The pounding of jogging can be a problem later in life if this is done on concrete or other hard surfaces. While the new designs in track shoes are helpful, jogging on a cinder track is still probably the best. If you are beginning exercising by jogging you should have a simple physician-monitored exercise test before you start. When you start to jog, remember that each person has a natural loping gait which is uniquely yours. Always use the "castle turret" (jog, walk, jog, walk, jog) method of starting to jog. Start with your natural jog, and when you develop moderate shortness of breath, slow the jog to a walk. After you recover your breath return to your natural jog. Keep doing this jog, walk, jog each day you practice. Three times a week every other day perhaps would be the best. You will find that gradually the jogging is longer and the walking is shorter until fairly soon you will be able to jog without the walking.

If you use indoor equipment obtain good instructions from a trainer. The same slow-fast training as in jogging can be used on a treadmill or bicycle if the equipment is controllable. Swimming is one of the best forms of exercise since it doesn't involve pounding. Water's support allows many muscles to exercise that might otherwise be too weak to start exercising without injury. Ozonated water is best for swimming, with the lowest chlorine possible. Use a snorkel and face mask to avoid chlorinated water.

The greatest stimulation to muscle growth is in fact the use of muscles right up to almost the maximum of the muscle's ability. If one goes past the muscle's ability you have injury which in the long run will impair the development of muscle growth. The greatest deterrent to muscle growth is lack of use. If you don't use it you lose it.

The Poisons You Know About

Alcohol

You know this subject well. Excess alcohol (ethyl alcohol) is an obvious problem and will not go away. Points that you may not know: The severity of alcohol as a problem depends on how you protect yourself from the potential free radical generation produced by the breakdown products of alcohol. Zinc is a critical component of alcohol's enzymatic breakdown. All of the antioxidants are necessary to protect the body. The percentage of the total calorie intake from alcohol versus the percentage of good food is critical. A person eating extremely good food in substantial amounts can tolerate with safety more alcohol than a person who is not eating properly. The mineral molybdenum is necessary to convert the sulfites in wine to the nontoxic sulfates.

Since some red wine (grape skin contains resveratrol which blocks tumor growth) has been shown to be beneficial to health. We do not tell you not to have alcohol in general. The distilled spirits usually contain some undesirable products and therefore are not recommended.

People who have been unable to control drinking in the past must not drink. Otherwise, one to two glasses of a good red dry wine may be helpful in increasing stomach secretions, the processing of food and the appetite.

Smoking

We must insist on elimination of smoking in public places. While leaded gasoline has been eliminated, some of our other standards of poison control have been relaxed too much. When you go to a restaurant the Maitre'd might as well ask you, "Do you want smoking or chain smoking?"

The weed tobacco has a predilection to pick up cadmium and arsenic from the soil. Cigarettes when burned produce a huge amount of free radicals which injure the bronchial and air sac lining. Tar breakdown products are carcinogenic. Nicotine is the addicting substance and is not as severe a poison as most people think. We look at a cigarette as a cadmium, arsenic and free radical delivery system which causes disaster. Cadmium is number seven on the poison list and interferes directly with the immunity system, displacing zinc. Arsenic is number one on the list and poisons the energy production of the cell.

You must not smoke and you should not allow anyone to smoke in your home. Secondary smoke is poisonous.

Step One in stopping smoking:

Wrap the cigarette pack in an 8 1/2 X 11 sheet of paper and place two rubber bands around it. When you want a cigarette you unwrap the pack, flatten the sheet out, place the cigarette to the side, rewrap the pack and replace the rubber bands. You may now smoke the cigarette.

Step Two:

Do not give a cigarette away or accept a cigarette from anyone. You don't want to poison your friends and you don't want your friends to poison you.

Step Three:

Buy one pack at a time.

Step Four:

Place wrapped pack in the trunk of your car.

Step Five:

If you can't quit on your own by this time, acupuncture or hypnosis can help. Quitting is a mind set and you must want to quit. Nicotine gum and sprays have limited value.

Cigarette smoking, depending on amounts smoked, reduces life expectancy by about six years.

A 100 pound woman smoking one pack of cigarettes a day is probably worse than a 200-pound man smoking 2 packs of cigarettes a day.

Women are 2 1/2 times more likely to smoke if they have dental mercury amalgam fillings than if they don't.

Antioxidant System Protection

The antioxidant system is the great protector of cell integrity and must be supported at all costs. Super Oxide Dismutase (SOD) is the fifth most common protein molecule in the body. Life expectancy is dependent on this protection from free radical damage.

Free radicals are caused by ionizing radiation such as x-rays, radioactive compounds, ultraviolet light, cosmic radiation, plus the intake of free radicals from breathing oxygen (hydroxyl free radicals).

When free radical generation in tissue is greater than the neutralizing effect of the antioxidant system then tissue damage occurs.

Not all free radical generation is detrimental. We could not live without it (see Cancer Section and Immunity). Avoid overloading the antioxident system. Protect it by following these guidelines:

*Avoid unnecessary x-rays, including mammography (see *thermography testing*).

*Avoid smoking (both active and passive). (see *poisoning*).

*Avoid excess alcohol (see section on *poisoning*).

*Avoid toxic metal poisoning (see *poisoning*).

*Avoid excess skin exposure to ultraviolet lights.

From a practical viewpoint it is important to understand that the antioxidants don't work by themselves. They work in a continuing cascade of reactions which needs every part of the chain in order to work. The chain includes carotinoids, vitamin C, vitamin E, selenium, zinc, copper, manganese, glutathione, vitamin B2, and niacin. When the research on vitamins is conducted in double-blind placebo controlled trials as with drugs, the study will be invalid unless the above mentioned components of the chain are included.

Hazards That You May Not Know About

Childhood Immunizations

In all health decisions one must weigh the benefits against the possible risks. We are seeing a huge increase in autism plus a spectrum of neurological diseases in children with many names, such as Attention Deficit Disorder. While dietary and genetics may be a factor, it is very clear that there is a close correlation between these conditions and immunization with agents that contain thimerosal as a preservative.

A pregnant mother with amalgam fillings in her mouth and/or excessive ingestion of mercury-contaminated fish transfers this poison to the growing fetus. The fetus becomes sensitized to mercury and at birth appears to be normal in all ways. Any immunization containing thimerosal produces a metal-induced autoimmunity reaction. Since mercury is primarily a nervous tissue poison, the impact of the autoimmune reaction is in the nervous system. Thimerosal, when injected into the body, immediately transforms into ethyl mercury, a very severe poison.

The risks of serious lifelong disabilities from the use of thimerosal containing vaccines outweighs the risk of the childhood diseases and tetanus. Do not allow your children to receive any immunizations which contain thimerosal.

Immunization Against Bioterrorism

In 1965, under the auspices of the Medical Service Foundation of America, I led a team of volunteers from the United States to the Dominican Republic with the purpose of instituting a smallpox vaccination program throughout the country. The Dominican Republic was chosen because of an almost total lack of smallpox protection. Starting the program, the very first vaccine I gave to Dominican Republic President Donald Reed Cabral with the handheld jet gun (dermajet).

This very efficient mass immunization tool, the dermajet, injects a metered dose of 0.1 ml of fluid intradermally. The smallpox vaccine came in multiple dose ampules designed for 100 people using individual, single use, vaccination needles. We discovered that when we took the 100 dose vial contents and mixed them in a 500 ml bottle of n. saline, we now had about 5000 doses of the vaccine for use in the dermajet. We found this method was basically painless. The metered dose using no needles almost completely eliminated any complications.

This method expands the supply of smallpox vaccine 50-fold. The U.S. now has adequate smallpox vaccine to immunize the entire country and I feel that the country should be prepared for mass immunization if documented cases of smallpox are found anywhere in the U.S.

During the plagues (bubonic, smallpox, and the influenza of 1918) the healthy people were usually reasonably resistant. This is another reason for all people to seek super health.

The Toxic Dentist

The most serious health hazards in the United States which you may not know about are the dental procedures which unfortunately have become standard.

While I have been practicing preventive oriented medicine for twenty-five years, I have to confess that I really didn't force patients to take their dental restorations seriously until about eight years ago. My amalgam fillings were replaced with composite about thirteen years ago because I knew it was the right thing to do. When I saw more and more dentists as patients with severe problems which were only solved by chelation (pulling out) of mercury, I realized I must address this poisoning in my patients.

Amalgam fillings are called silver fillings (I guess they look silvery), but they are made of 50% mercury, 35% silver, 15% zinc and other metals. The powdered metals are in solutions with the mercury which then leaches out every time one chews. About 80% of this poisonous vapor is absorbed and goes directly into nervous tissue. Mercury is number three on the EPA hazardous list and as far as we are concerned it is the most serious poison in the group due to the extreme toxicity plus chronic exposure. Untreated the half life of mercury in the brain is about 20 years. Mercury tends to attach to the sulfhydril groups which are central to the functioning of the nervous tissue. Mercury winds up poisoning the midbrain which controls all of the body's automatic functions. The hypothalamus (which regulates the pituitary) is interfered with and therefore growth hormone, thyroid hormone, sex hormones, and adrenal hormones are out of balance. The huge number of symptoms created by this interference include chronic fatigue, memory loss, irritability, constipation, etc., etc, causing Alzheimer's, multiple sclerosis, and other neurological diseases.

When amalgam fillings are removed by mercury-free dentists, we chelate the metal out of the tissues at the same time and we see a huge difference in the functioning of the patient. Proper supplementation is also needed at that time.

We see many mouths full of both amalgam fillings and gold next to each other. These dissimilar metals produce galvanic currents which disturb the brain as well as the pituitary. We need tooth restoration materials which are biocompatible and also non-conductors of electricity.

When choosing a dentist one should call the dentist's office and ask "Do you place amalgam fillings in anybody?" If the answer is yes, say "thank you," and call another dental office. You must go to a mercury-free dentist.

Another potentially hazardous procedure is root canal therapy. It is noble to try to save a tooth; however, it shouldn't be done or it may produce debilitating disease. There are about three miles of dentin tubules in a molar and these come out at right angles to the root canal which supplies the nerve innervation and blood supply to the tooth. If there was not complete sterilization at the time of the root canal surgery, anaerobic bacteria will live in these tooth tubules and secrete botulism-type neurotoxins.

Boyd Healy, Ph.D., has developed a test to determine the level of root canal toxins. He states that about 1/4 of the levels are nontoxic, 1/2 are toxic and 1/4 lethal in the sense that it will be a major factor in the death of the patient.

Chronic fatigue, chronic pains in various locations, neurological conditions etc.,—all can be caused by these toxins. Evidentally there are some new techniques being developed which can eliminate the toxic root canal. These procedures involve laser sterilization and the use of Biocalex/MTA in root canals.

The use of fluoride in dentistry is another potential pitfall. The use of fluoride has almost become the "holy grail" of eliminating dental caries over the years. It would appear that we have been tricked by the fluoride industry to pay for a toxic waste that they had trouble disposing. Despite all of the hype, there is really no evidence that fluoride is anything more than a poison. To quote Dean Burk, Chief Chemist Emeritus USNCI: "Fluoride causes more human cancer death and causes it faster, than any other chemical."

Studies show that the more fluoride a child drank the more cavities appeared in the teeth. Recent studies show a correlation between exposure to fluoride and the lowering of IQ in children. Bones treated with fluoride may become more dense, but also become more brittle.

We should all refuse any fluoride treatments in dental offices, fight local politics for removal of fluoride from the water system, and avoid all toothpaste containing fluoride and sodium lauryl sulfate.

The Toxic Physician

The second most serious health hazard is the toxic physician who is the poisoner. Without the prescription pad most physicians would feel impotent. In our office, we probably use about 10% of the drugs we used in previous years. If we could find a natural nontoxic substitute we could eliminate the 10% we presently use.

The huge amounts of antibiotics in use today are not needed. If the person is de-poisoned and eating properly, usually an herbal immune stimulant is all that most mild infections need (viral or bacterial).

Just as the mercury story is the disaster for dentistry, cholesterol is medicine's Achille's heel. The cholesterol theory of vascular disease is just not true. We don't need statin drugs for cholesterol control (see Heart Disease in Troubleshooting).

The tranquilizer drugs are mostly stop-gap measures for the stress of our lives. When we have proper nutrition and are depoisoned we don't need them.

The chemotherapy drugs as presently used are almost all disastrous and I predict will be discontinued over this next decade. They basically don't work (see Cancer in Troubleshooting).

Standard medicine is crisis oriented and its finest work is in that field. The ability to save peoples' lives after serious accidents is often remarkable. Where standard medicine fails is in the treatment of degenerative diseases which we generally associate with aging. The drugs are basically blocking agents, the breakdown products of which cause all kinds of symptoms. Prescribe enough drugs and you have an excellent chance of drug incompatabilities.

Modern medicine and good medicine do not necessarily equate. Note whether your physician always gives you a new prescription each time you visit. A good physician is a teacher who limits the drug usage to the least amount compatible for your condition. You and only you are in charge of your health.

Maintenance Testing and Interpretation

We wish that God would send us a certified perfect man and woman. Then we could do all of the testing needed to find out what ideal parameters for each test would be. Perhaps they might look like the man and woman on the cover. Right now we are stuck with what standard medicine has done for standards and frequently they are not helpful. The group of people making up a test standard include the healthy, the sick, and the walking wounded. Then you find the mean of that test level and declare that two Standard Deviations on each side are normal. When dealing with an increasingly poisoned population that are eating fast food you know that the parameters may be "normal" but are not even close to ideal.

Example: Women live longer than men by approximately 5 1/2 years. This difference is due to over 30 years of menstrual cycles which keep the iron stores low. Iron is the catalyst of free radical reactions. Once women reach menopause and start retaining iron (as men have during their entire life), they join the ranks of males in increasing incidence of vascular disease. A 1992 Finnish study showed men with ferritins over 200 had 2 1/2 times the heart attack level of men with ferritins below 200. Despite this, normal ferritin ranges of men are listed up to 275. We feel ideal ferritin level is about 40-50.

Before laboratory testing, it is vital to have a complete history and physical exam as indicated. Physician observations are an important adjunct to proper diagnosis. A detailed patient history is probably the most important part of the examination and should include a dietary history. Accurate patient participation is as vital as a physician who is a good listener and questioner.

The first time we do testing on a new patient we recommend rather extensive biochemical testing. We are looking for early trends in areas where prevention-type therapies can have a major effect on the health. We try to keep people out of hospitals by finding areas of correctible problems before they show up as disease. This is maintenance at its best.

Test Categories

1. Standard testing: blood counts, chemistries, fat panels, electrolytes, liver panel, basic metabolic panel.

2. Hormonal testing: ILGF-1, DHEA-s, thyroid hormones, sex hormones, PSA.

3. Mineral testing: Hair analysis is the most accurate and cost-effective screening method for toxic as well as nutritional minerals. The EPA recommends this test for both arsenic and mercury. In addition, checking urine metals before and after a treatment with a mineral chelator can evaluate mineral status. Whole blood metals can be determined to accuracy and parts per billion. However, many minerals are found only in tissue and may not be found in blood.

4. Immunity testing: Immunoglobulins help direct us to immunity defects and possible allergies. A basic lymphocyte profile can identify the immunity system componenet that controls cancer. Natural killer cells and toxic lymphocytes increase when immune stimulants are added and poisons are removed. Tests include the pathogenic yeast Candida Albicans..

5. Essential fatty acids: are the precursors to vital regulatory proteins called prostaglandins.

6. Stomach and pancreatic enzymes: are tested by a "James Bond" type of test. An FM radio transmitter in small capsule form

is swallowed by the patient and information is returned by radio signal analysis.

7. Inflammatory disease indicators: blood ferritin, lipoprotein (a), c-reactive protein (high sensitivity), homocysteine, TAG, total antioxidant gap.

8. MELISA testing: for hypersensitivity to metals causing autoimmune diseases.

9. Optional ICG food allergy and IGE respiratory testing.

10. Breast thermography: Thermography is a form of diagnostic imaging based on infrared heat emissions from targeted regions of the body. This is infrared photography of the breast heat patterns. Thermography has been developed over the past thirty-five years in major medical universities and women's clinics throughout the world. It is the method of choice in screening for breast disease. It has an 88% sensitivity with as much as 96% reliability for indicating cancer, especially in premenopausal women. Thermography offers a very early warning system, often able to indicate a cancer process five to eight years before it would be detectable by mammography.

As the body's cells go through their metabolic energy conversion processes, they emit heat. Skin temperature is a reflection of the quality of blood flow in that area. Thermography is able to register these heat emissions, display them on a computer monitor, and thereby provide a diagnostic window into the functional physiologic status of a given body area. As tumors form, they develop new, abnormal blood vessels called *neoangiogenic vessels*. Thermography excels at detecting these vessels. The thermography procedure is simple and noninvasive. No rays of any kind enter the patient's body and there is no pain or compression of the breasts. Never do a test which increases the disease you are testing for. Mammography fits this test description. Thermography is quite useful and should be substituted for mammography. At present, there are approximately 1,000 thermography devices in the United States for providing this detailed, clinically valuable information.

Supplementation

Our diet at the present time is made up of many highly processed foods and contains so many poisons that we need help in protecting our bodies even more in this century than at any time in the 20th Century.

When a "health professional" says that supplements make expensive urine and that you don't need any supplements just remember that about 1/2 of the population takes less than what the RDA level of vitamins recommended. That standard is basically the minimum needed to prevent deficiency diseases if your metabolic machinery was perfect and you weren't poisoned. (See the appendix for a list of 275 poisons found in the workplace).

Nutrisure OTC was created to supplement a minimum of what the average "generic person" needs daily. Two Nutrisure OTC with each meal could be a reasonable start toward better health. For a smaller person (about 100 pounds) two tablets twice a day may be satisfactory. We can do much better than this when we suggest supplements on an individual basis following the results of the patient's testing.

When one is using melatonin at night, a sublingual tablet, Bevitamel, containing 3 mgm melatonin, vitamin B12 (1000 mgm), and 400 mgm Folic acid is recommended. Generally 1/2 to 1 tablet under the tongue is recommended. See the Westlake Laboratories' listing in Physician's Desk Reference under Westlake Laboratories. These products may be obtained from Westlake Laboratories 1-(888)-WSTLAKE (978-5253).

Nutrisure OTC

Two each meal for average adult, six per day, contain

Beta carotene	15,000 IU
Vitamin A	10,000 IU
Vitamin C	1,200 mgm
Vitamin D-3	100 IU
Vitamin E	400 IU
Thiamine Hcl	100 mgm
Riboflavin	50 mgm
Vitamin B6	100 mgm
Folic Acid	800 mcg
Vitamin B12	100 mcg
Biotin	300 mcg
Pantothenic Acid	300 mcg
Calcium	500 mgm
Magnesium	500 mgm
Iodine	200 mcg
Zinc	24 mcg
Selenium	300 mcg
Copper	1.8 mgm
Manganese	18 mgm
Chromium	300 mcg
Molybdenum	100 mcg
Potassium	110 mgm
Choline	70 mgm
Inositol	100 mgm
Bioflavanoids	100 mgm
PABA	50 mgm
Vanadium	50 mcg
Boron	1.5 mgm
Trace elements	34 mcg
L cysteine/NAC	200 mgm
L methionin	12 mgm
Glutamic acid	24 mgm
Betaine	48 mgm

❖ ❖ ❖

Energy Medicine

Stanley M. Gardner, M.D.

In order for a diseased body (a body in disharmony) to heal itself, there must be a highly sophisticated and orderly set of commands that cause physiological activities to take place. Several systems, including enzymes, blood clotting, hormones, prostaglandins, and immunity all work together to regulate these activities. Communication between cells takes place at the energy level.

We can measure effectiveness of some of these systems by electrical measuring devices, such as electrocardiogram (EKG) for the heart, electroencephalogram (EEG) for the brain, and electromyogram (EMG) to measure conduction in skeletal muscles. The results of this testing tell us if the electrical (or magnetic) energy system is intact or blocked in some way. When the energy flow is blocked where tissue is damaged, abnormal EKGs, EEGs, or EMGs are found.

Standard medicine also presently utilizes magnetic resonance imaging (MRI), ultrasound and thermography for diagnostic purposes. Energy is widely utilized within standard medicine for treatment purposes. Pulsed electromagnetic fields (PEMF) are a low-energy, low frequency current that is used to stimulate bone growth in non-union fractures. This energy stimulates fracture healing in 80% of fractures that have shown no evidence of healing during the previous 6-12 months of stabilization. Sports injury therapists use ultrasound, hot and cold packs, electricity and magnetics in an effort to reduce damage and facilitate healing in injured tissue. Light therapy is widely used in depression, especially Seasonal Affective Disorder (SAD). Art therapy, music therapy and aromatherapy are merely different vibrational frequencies widely used in psychological care. Radiation therapy and laser usage are additional therapeutic tools with tightly controlled vibrational (energy) frequencies.

Different cells and different organs respond to different vibrational frequencies. Sisken and Walker reported in 1995 the following ideal frequencies that stimulate growth and repair:

Frequency	Affects
2Hz	Nerve regeneration Neurite outgrowth from cultured ganglia
7Hz	Bone
10Hz	Ligament
15, 20, 72Hz	Skin necrosis Capillary formation Fibroblast proliferation
25 & 50Hz	Nerve growth factors

When a vibrational frequency is externally applied to tissue that matches its individual ideal repair frequency, this accelerates the cascade of activities that trigger healing. These frequencies can also open up the lines of communication with surrounding tissue or even the brain to direct these processes.

Energy medicine takes the concept of vibrational medicine one step further. It maintains that living matter possesses some kind of "life force" or "life energy" that harmonizes with the known laws of science. These are energy flows that build and then maintain our physical body, like an energy template or blueprint. These forces unify and integrate the body in perfect harmony.

There are various life experiences that interfere with these natural flows and cause imbalance. If imbalance continues long enough, disease may eventually occur. This may manifest itself as pain, stiffness, swelling, or tiredness, among other things. Eventually organ breakdown may take place at the digestive, respiratory, reproductive, vascular, or nervous system level.

There are several ways in which blockages to energy flow arise. These encompass sources from within and from without.

1. Diet: Energy is derived from the soil. The closer we get to raw natural foods, the more complete energy is brought into our body. As food is processed and refined, energy is reduced or eliminated, to the point where destructive energy could be ingested.

2. Mental/ Emotional: The "attitudes" of worry, fear, anger, sadness and pretense directly affect energy pathways. The acronym FEAR describes how these attitudes often arise--False Evidence Appearing Real.

3. Working habits that are frantic and hectic, void of peace, interfere with energy flows.

4. Physical accidents

5. Poison ingestion: Heavy metals, including arsenic, mercury, lead, aluminum and cadmium clearly interfere with energy transmission within the body. Pesticides and various air pollutions and gases also need to be considered in this list.

6. Hereditary characteristics

7. Weather conditions

If we are to bring harmony back into our body and life, we must address each of these issues that apply. Additionally, there are various tools for relaxing the body and removing blockages. We will mention two, breathing and therapeutic touch.

Breathing represents another source of energy. It begins with exhaling, so that we can completely expel old breath and bring new breath, or new life, into the body. During the exhale we clear the body, emotions and mind, so we are ready to receive all of the levels of life energy for all the body's functions.

In therapeutic touch, the hands are used to channel the energy around us. One method of therapeutic touch, Jin Shin Jyutsu, is an ancient Japanese study of what the human being is from nine month development in utero through to the formation of all levels of the body from skin surface to bone. It explores all of the levels of man: energy, life, physical, emotional, and spiritual.As a healing art it recognizes the body's (person's) innate potential to heal itself. It involves gentle application of hands to the body to help all of the different levels of life energy clear.

There are many other Eastern healing arts that have been extremely useful for health, used for thousands of years. Acupuncture from China is the most widely known. This utilizes placement of needles in certain places along flow lines called meridians to stimulate energy (body's life energy, "chi," "Qi," or prana in other languages). Reiki, Qi Gong, and yoga are a partial list of philosophies that utilize energy in the healing process.

Chiropractic, osteopathy and structural integration (rolfing) add a component of manipulation and deep tissue work to the energy picture. They recognize the importance of a stable body balance, and the compensatory changes that take place within the body with misalignment (not to mention the blocked energy caused by misalignment). Massage is a combination of hands-on pressure techniques which are designed to increase relaxation and body harmony by removing energy blockages; ie, muscle spasms and lymphatic stagnation.

Homeopathy is based on the philosophy that the human body contains an electromagnetic control network, mediated through the water system. Certain vibrational frequencies from specific substances are "successed" into the water. The water then becomes the carrier of the vibration, without the need for the original substance.

There are now machines that assist in measuring which nutrients, medications or herbs will have the greatest impact on the body, based on the vibrational frequencies of the substance relative to the vibrational state of the individual. Kinesiology can also assist in determining strength or weakness to muscle groups caused by various substances.

All problems with the human body are energy related. Once the proper energy flows to and through the tissue, the body begins to function properly, and disease departs, leaving peace, harmony, and health, or wholeness, in its place.

The Economics of Maintenance

Is the so-called "health care system" interested in health? For years health care insurance basically paid the bills for disease, added a percentage and then increased the premium to cover any losses. Only when the premium became too much for the population to pay were limitations placed on the benefits. Since the insurance was for disease care and those payments were being reduced there certainly wasn't anything available for almost any type of preventive therapies. Medicare took the same track as the private insurance companies.

It would seem that the best thing you can do for your country after you reach medicare age would be to drop dead. Prevention of disease and extending life are counterproductive to the economics of all health insurance as long as they are controlled by crisis medicine mindset.

The life extension insurance of the future will involve a combination of a high value (say $1,000,000) life insurance policy with a mandatory preventive oriented health plan. What the insurance company would save each year from not paying out the death benefits would buy a lot of prevention. It would pay for a lot of valuable therapies not covered now that would be substituted for more dangerous standard procedures. Too often I have heard, "I'm going to have bypass surgery since it is covered by insurance. Your alternative procedure is not covered." Who is making the decisions of importance regarding your health?

It All Fits Together

The president of ABC Company goes to his business consultant and says, "My company is going bankrupt. What is the single reason ABC Company is going bankrupt?" After inspection the consultant brings a report back to the president. You have 100 employees with 5 embezzlers. You pay too much for raw materials, too much for lawyers, your quality control is bad, etc., etc. Fifty things are wrong with the company.

If you want to know a single thing causing your bankruptcy it is because you are president. Good health is only obtained when you are in charge of your body and doing things right.

The Future of Medicine

The drug industry controls medicine and to some extent surgery. Medicine lobbies to Congress are omnipresent and powerful. It scares me to think that drug coverage would be covered by medicare because it is a right (like the Bill of Rights). This means that the only safe investment in this country is investing in the drug companies.

If drugs solved basic problems I would go along with this. We must somehow change disease care to actual health care and do it in a more affordable manner. I feel very fortunate to have spent the 25 years in standard medicine and the last 25 years in preventive-oriented medicine. I feel sorry for the average physician who is beginning to realize that he is working with an unworkable "health" system.

III. TROUBLESHOOTING

Introduction

If one follows the general maintenance rules after having some understanding of how the body operates one should have excellent health and probably not need a crisis care physician. All troubleshooting starts with a review of proper maintenance as it is tailored for the individual and how closely the plan has been followed.

Symptoms are feelings that things are not what they should be. Signs are visual observations of something wrong. Testing is trying to diagnose by looking at non-obvious signs. When a car's engine "knocks" it could be many things but low octane fuel might easily be the answer. When you find the proper answer usually the many symptoms and signs go away very rapidly.

I had a patient come in saying, "I have five complaints." We did our usual workup and changed diet, gave her needed supplements, de-poisoned her, etc. After about three months she returned and said, "I hope you're not angry with me but I gave you five symptoms of my list of 25. I have been kicked out of so many doctor's offices with a prescription for tranquilizers that I decided to try something different with you. What is interesting is that four of the five original list of symptoms are gone. Of the list of twenty that I didn't tell you about, sixteen symptoms are gone and now we are back to five complaints." I explained that we were trying to have the body repair itself and we weren't treating symptoms. It's interesting that we really didn't do any better on the symptoms that we knew about compared to the list we didn't know about (80% reduction in symptoms).

When we discuss various disease groups in this troubleshooting section we have to talk in generalities because of the huge number of diseases and syndromes. Please note that it has been estimated that there are over 250,000 diseases and syndromes known to mankind. Many of these are devastating and many are not associated with decreased longevity or any disability. Infective agents such as tuberculosis can be in many organs and yet introduce themselves as different diseases. Tuberculosis of the lungs, skin, adrenals, bones, etc., all start with very diverse symptoms and each is treated by different specialists.

Degenerative disease death is caused basically by loss of cell energy following free radical damage to cell wall integrity. If this injury is to blood vessel cells you have vascular disease, if it is to the cell nucleus you might have cancer, if it is to nerves and brain cells the result might be neurological diseases, etc.

Standard medicine is very good at making diagnoses of most diseases. A set of symptoms placed together may be called "ABC Syndrome," etc. We are frequently asked to give a "second" opinion. The patient says, " I'm on five drugs from three doctors. Can you help me?" I say, "Do the three doctors talk to one another? Have they asked about nutrition? Have they investigated poisonings?" When the answers to these questions are no, I feel that we can help the patient.

We have been taught in medical school to kill the disease. Upon the death of the disease good health will arise somehow because good health is a lack of diagnosable disease.

The body is remarkable in its recuperative powers when nourished, de-stressed, and de-poisoned. We cannot discuss all diseases and conditions. However, the main diseases will be discussed and some individual comments on some of the common problems which you can fix.

Heart disease and vascular disease are now the cause of death in about 45% of US deaths (down from 55% about 20 years ago). Cancer and immunity deficient diseases are about 40% of the deaths (up from about 30% 20 years ago). These two still add up to 85%, 15% for all other kinds of death.

Lemon Law

Since you are alive and you can't send your body back for any recall we must try to correct any defects with which you might have been born. Sometimes that's a very big job that only you can do.

It has been said that the best thing you can do is choose good parents. I'm not sure how you do this, but genetics is very important in the development of problems at any age. Picking good parents basically means parents that also have applied the correct health principles in the past.

If an experimental animal during pregnancy is made zinc deficient or is poisoned with cadmium, the current litter and the next two litters of offspring will be immune deficient despite proper nutrition. Example: Folic acid deficiency during pregnancy can produce neural tube defects (spina bifida).

A baby is a malignancy as far as the mother's body is concerned. This self-limited tumor grows at a rapid rate and takes out of the mother what it wants and needs. You want the baby to have all of its "marbles," so to speak, when the baby is born. The mother must replace all of the components that the baby needed during pregnancy. If she becomes pregnant again before the nutrition is replaced the next fetus is at risk for defects or she is at risk for disease.

Everyone has an "Achilles Heel" since no one is born perfect. Even identical twins, while very close, are not exactly the same, since after the fertilized egg separates to produce "identical twins," external forces throughout life are different for the twins.

While genetics are important and new "genes" for some disease are being found daily in the newspaper, the principles held in this manual are still very useful if carefully followed. When a defect is found, frequently more attention paid to some particular nutrient combination and/or de-poisoning can keep detrimental defects which affect quality of life to a minimum.

Heart Disease and Vascular Disease

Congenital heart disease is so variable that it is beyond the scope of this book. <u>Some</u> congenital problems are solved by innovative surgery, but prevention is <u>still</u> the most important solution to congenital heart disease.

Heart disease is part of the degenerative diseases to which we are very susceptible. It is Coronary Artery Disease (CAD) with which we are most concerned. CAD can certainly lead to heart attacks (Myocardial Infarction, MI) and is the most common cause of sudden deaths.

In my first 22 years of my medical practice I was a family physician with a special interest in exercise cardiology. I am a member of the American College of Cardiology and at a meeting in New York City in 1975 my entire professional life was changed. I was sitting next to a doctor who told me that he was taking EDTA chelation therapy for CAD. I said, "Do you have lead poisoning?" He said, "no; however, it works well against the disease." I frankly didn't believe him because I had never heard about it before. I had sent many CAD patients for CABG (Coronary Artery Bypass Graft) surgery. Two scenarios had developed from patients returning from surgery. Some, of course, didn't recover from the surgery, but that was expected since there was no other treatment. One group praised me for the diagnosis and "proper" treatment by the surgical referral. I was proud to be part of the team. Another group said, "This man attacked me with a chain saw and I feel terrible. I thought that you were my friend." At that point I wanted to join another team.

This doctor suggested I go see Harold Harper, M.D., in Hollywood, California, if I wanted to learn more about chelation therapy. I was going to California for another meeting and I thought I would stop at Dr. Harper's office to see medical quackery in action. Dr. Harper greeted me with, "Welcome to the land of fruits and nuts!" He said, "Don't believe me but talk to my patients." He was right. I didn't believe him. It was a fascinating experience, as I intended to spend three hours. I spent three days with him, learning all I could. At least 90% of his CAD patients were obtaining excellent improvement from the intravenous bottles which took about 3 hours, two times a week for about 20 treatments, followed by less frequent treatments for six or more months. With nutrition and vitamin therapy this EDTA chelation was doing far better than CABG surgery with no mortality and at far less cost. I brought the therapy back to Cleveland, Ohio, and both I and my patients have been extremely happy with the results.

The 250,000 miles of tubing we have in our body are all treated with chelation therapy, not just a few inches of the area as treated by the CABG operation. On a cost accounted basis, $50,000 paid to treat a few inches of a 250,000 mile problem represents more than all of circulating money in the world per patient (I realize this is trivial pursuit; however, it is food for thought).

Besides the heart, chelation improves the circulation in the cerebral arteries as well as the peripheral leg arteries.

EDTA therapy is the repeated intravenous use of EDTA with magnesium. The therapy stimulates a number of actions at once which have a profound effect on vascular repair. Calcium is pulled out of vessels where it is interfering with cell energy in the heart and vessels. It is replaced in bone through pulsitile parathyroid stimulation. Heavy metals are removed from the vessels which restore normal small vessel functions.

The blood vessels of any size have blood vessels that supply the nutrients needed for vessel repair. We call these *Vaso Vasorum*—VV (The Vessels of the Vessels). Vascular disease and especially catastrophic disease such as rapid blockage of a major coronary vessel have been shown to be inflammatory reactions. The loss of antioxidant function at that point is the major reason for the reaction. Chelation therapy's most profound effect occurs in the VV of the vessels and its effect on reducing platelets' stickiness at the disease site.

When heart muscle cells do not receive adequate oxygen and nutrients, the ATP production in the mitochondria goes down. If this goes to a certain level where the calcium pumps are not functioning, calcium ions enter the cell and poison mitochondrial function. Dr. C. F. Peng, et. al, in 1977 showed that EDTA removes the calcium from the nonfunctioning mitochondria and restores the mitochondria to full energy production. Cell energy is central to health.

The heart muscle is designed to pump almost forever--certainly longer than just about any other part can function in our body. Unfortunately the limiting factor is coronary artery delivery of oxygen and nutrients. The pump can also be poisoned by heavy metals. In May 1999 it was shown that Idiopathic (means "don't know the cause") Dilated Cardiomyopathy (heart muscle pump failure) occurred when arsenic was 250x, antimony was 12,000x and mercury 22,000x higher than normal respectively. Idiopathic Dilated Cardiomyopathy is caused by heavy metal poisoning.

EDTA therapy is still a standard approved and safe toxic metal chelator which binds these poisons and removes them from the body by way of the urinary system. Arsenic, lead, mercury, cadmium, nickel, antimony, aluminum, iron, uranium, plus many more are removed from the body by EDTA chelation therapy. We feel that this is a major factor in the success of the therapy.

The coronary angiogram is a procedure that has become the diagnostic gold standard of cardiology as far as coronary artery disease is concerned. Non invasive testing prior to the angiography is often considered inadequate by cardiologists. The use of angiography has escalated, despite the fact that Dr. Carl White, et al, in 1984 showed that the physiologic effects of the majority of coronary obstructions cannot be determined accurately by conventional angiographic approaches. This procedure usually leads to the next stop in the path toward the recommendation of angioplasty or CABG surgery.

A standard cardiologist, Thomas Grayboys, M.D., who does second opinion work in the Boston area found that 80% of the patients who were told that they critically needed invasive interventions did not need the procedure by standard medicine's own criterion.

Cardiac function as a pump is the most critical parameter for determining the magnitude of the CAD. It is not the number of diseased vessels that predicts outcome. Chelation therapy should be tried first and other interventions considered only in the rare case of failure of chelation.

Complications from chelation are basically nonexistent when the procedure is done properly. One problem with chelation therapy is that it is a good remover of important minerals such as zinc and manganese. It is very important that these mineral levels be monitored during chelation therapy and replaced as indicated during the therapy. Good nutrition is critical.

Complications from CABG are considerable without adding the possibility of death directly from the surgery. Reported in the New England Journal of Medicine this year was a long term assessment of cognitive function following CABG surgery. A decrease in cognitive function of 53% was noted at discharge and 42% remaining at five years after surgery.

Since chelation improves small vessels throughout the body, including the brain, we note an improvement in the thinking of most patients who have chelation therapy. Further discussion in detail can be found in the Appendix.

Treatment of Cancer in General

No generic recommendations for cancer treatment can be made. The diagnosis is important but only represents the fact that a biopsy piece of tissue was obtained and that the biopsy was taken from an area where cancerous cells were able to accumulate enough in one spot to make a diagnosis. The handling of the biopsy material is important and frequent mistakes are made if the biopsy is mishandled. A positive diagnosis is not always correct and certainly a negative biopsy does not rule out cancer.

1. Cancer is not a surgical disease. It is a metabolic process disease. The primary tumor removal, if able to be done by surgery without jeopardizing the life of the patient, is probably a good idea. This is called "debulking." The lymph nodes and/or any other metastases should be left undisturbed by the surgeon.

2. With rare exception, standard chemotherapy should not even be considered. This does not increase quality or quantity of life.

3. Radiation therapy in far advanced cancers can be considered for reducing tumor size that is pressing on a life threatening organ, nerve, ducts or other vessels. Except for these conditions radiation is not a cancer therapy since it injures too much normal tissue.

4. All drugs taken by the patient should be carefully evaluated and stopped if not absolutely needed.

5. Under the direction of a preventive-oriented physician, the patient's life style must be changed for truly long term survival. Help and guidance in stress reduction techniques, individualized diet and nutritional needs, reduction of poisonings, and enhancement of the immunity system must all be implemented.

6. The Lentz filtration method for metastatic cancer treatment will be the best method and should be tried when it becomes available. This is the technique which filters the solubilized tumor necrosis factor receptors out of the blood. The immunity cells are no longer blocked and are then free to do their job killing individual cancer cells. It is in its final approval process with the FDA.

7. Insulin Potentiated Therapy, or IPT, is a relatively new method of cancer treatment which may well be a satisfactory treatment for advanced disease. While I suggested not to use chemotherapy, this treatment uses some standard chemotherapy in only about 10% of the standard dose. Cancer cells have 16 times more insulin-like receptors than normal cells and are huge sugar eaters. Cancer cells and normal cells are similar; however this difference in number of insulin-like receptors is a huge difference between the cells. This therapy employs the use of insulin to deliberately reduce blood sugar. During that state introduce low-dose chemotherapy appropriate for that particular cancer. The cancer cell wall is very susceptible when in a low blood sugar state taking up a large portion of the chemotherapy. The dose is not high enough to injure normal cells. The patient remains free of the toxic symptoms and changes in body functioning that are produced by standard chemotherapy. There is no loss of hair, but the most important part is that the dose is not enough to injure the immunity system.

The term *palliation* is used if whatever treatment that is given is used with the idea that the patient will die from the disease. It has long been a "noble" effort of the physician to relieve pain and suffering. The drugs used for pain relief unfortunately also kill the appetite. The continued eating of good food is critical to the continued life of the patient. Make sure that the goals of the physician are the same as your goals. You may well need a second, third, or fourth opinion so that you make the right decision.

Cancer

A 35-year-old woman went to her doctor with a golf ball sized mass in her breast which the doctor diagnosed as cancer. The physician said, "If you don't have that breast removed by surgery you will be dead in six months." The woman said, "I'm not going to have any surgery," and left the doctor's office. She had for years been on what I call a "Twinkie diet" which was so bad that she was lucky to live to the age of 25. She went by a bookstore and picked up a book on nutrition and cancer. She totally changed her diet and felt much better. She went to the same doctor one year later and was examined again. He said, "If you don't have your breast off now you will be dead in six months." She responded with, "That is what you told me one year ago." She left, not wanting surgery. One year later the same physician repeated the performance and she responded with, "That is what you told me two years ago and one year ago." The physician made a very interesting response to all of this, saying, "You don't follow my advice, so get out of my office!" One year later this patient arrived at my office and told me her story. We helped her with her problem for a while; however, she got into an accident and injured the breast, which bled severely. She went to the hospital and two units of blood were ordered. Somehow the blood was incompatible and she promptly died in the hospital. She did not die of cancer. Why do I relate this terrible story? Cancer is a chronic degenerative disease which responds to nutrition. It comes on very slowly. She probably had her disease at least six to seven years before the lump was found. Treatment should not be based on the attitude that "yesterday we could have gotten it in time but today it is too late." This attitude can panic patients and is counter-productive to the body's repair system. Part of the treatment of cancer is to remove stress from the body. Chronic fear suppresses the immunity system and the power of the physician was used as a negative hex (probably worse than the standard witch doctor). The preventive-oriented physician has to use all of his positivc powers to "exorcise" what a physician did to this woman.

The use of surgery to make a diagnosis and debulk the tumor is probably very appropriate. Debulking (reducing tumor load on the immunity system) is helpful since there is less tumor to be removed by the patient's immunity system. Cancer is not a surgical disease if it has spread at all past its original location. While standard medical care is irradiation and chemotherapy, these modalities (with a few exceptions) do not work. The susceptibility to these treatments by both the cancer and the normal cells is about the same. If you want to kill some crows in the orchard the shotguns will kill some crows, some song birds, some apples, and some neighbors.

Cancer is probably incurable without finding and correcting the cause. The cause is usually three in nature: stress, poor nutrition, and poisonings. If I had a magic wand and passed it over a person with cancer and the cancer went away we can expect that cancer or another one to start if we have not addressed the causes.

In our workup for cancer patients we try to find the causes. Correct the cause and the stress involved, stimulate the immunity system, and, depending on what type of tumor is present, institute other modalities. If the tumor develops and spreads to other areas in the body we call those areas metastases. The tumors learn to survive by spitting out the Tumor Necrosis Factor Receptors (TNFR). Those TNFR when solubolized (STNFR) circulate in the bloodstream and adhere themselves to the probes of the natural killer cells. These cells and others (toxic lymphocytes) have controlled cancer cells throughout your life up until now. When the cancer releases these receptors in large quantity the natural cancer killing cells are substantially impaired in their killing ability.

New techniques have been developed to filter out these circulating cell receptors and when they are greatly reduced in numbers these natural killer cells have a renewed ability to destroy cancer no matter where it is. These metastatic cancers can be killed very efficiently without damaging normal cells. The physician using this technique has to be careful that he or she doesn't destroy the cancer too rapidly. While I consider this the cancer therapy of the 21st century, it needs a healthy immune system to work well. We have to improve the immune system so that it can return to doing its normal surveillance of cancer cells.

Cancer is a process which is in all people and which is controlled by our immunity system at all times. Until we realize that cancer cells are with us continuously, will we be able to give up our "kill the cancer" philosophy?

Sexual Dysfunction

In both men and women libido is both psychological and hormonal. In both sexes testosterone is critical. While the woman has only about 10% of the level of the man's testosterone that is an important 10%. Giving a female dose of testosterone can be very helpful to the woman.

The preparation for intercourse from the brain's standpoint involves parasympathetic stimulation. This produces an increase in vaginal lubricating secretions in the woman and a penile erection in the man. In the normal progress of intercourse leading to a climax there is a swing in the autonomic nervous system control from parasympathetic to sympathetic. The shortness of breath created is similar to the sympathetic response of running. The greater the swing from parasympathetic to sympathetic accounts for the variation in the level of pleasure created at the peak of the orgasm.

If one has a lot of personal stress going on at the time this creates sympathetic stimulations and the parasympathetic mode the person tries to achieve may make climax impossible or less than satisfactory. While we may not be able to eliminate the interfering stress, the body can be made to tolerate it much better. Poisonings and malnutritional deficiencies are stressors which stimulate the sympathetic nervous system. Correcting these can frequently do more for sexual dysfunction than any advertised pill.

When the controlling hormoncs of the body are evaluated and balanced, and diet and mineral balance corrected, most sexual dysfunctions can be corrected.

Osteoporosis

Actually men and women have an equal amount of osteoporosis. You say that isn't possible! Women's bone structure is much lighter that that of the males'. When calcium goes out from the males' bones it winds up in the blood vessels and produces heart attacks and strokes, etc. Fracture is generally not a problem in the males until in the 80's. The fracture problem of osteoporosis in women starts rather rapidly as the estrogen goes down and the FSH goes up. FSH has been shown to directly interfere with the remodeling of bones and therefore anything to keep the FSH down is helpful. This is, or course, restoring the estrogen so the pituitary will stop producing the FSH. See Ostopeorosis Prevention under Maintenance.

Hypertension

Everyone agrees that hypertension is a serious risk to general health. We have about 250,000 miles of tubing in our body (yes, that is the distance from here to the moon!) and we need a certain amount of blood pressure to send blood through that huge network of mostly tiny vessels. If that pressure is too great it places undue stress on the whole circulatory system as well as increases the work load for the heart.

If the blood pressure is too low we obviously have an insufficient perfusion of nutrients and oxygen to our hundred trillion cells that make up the body. When there is insufficient perfusion plus red blood cell clumping and platelet aggregation we can have major areas of microcirculatory stagnation. I call it *gridlock*. This is the precursor to those muscle spasms ("charley horses") that can give trouble to an otherwise comfortable sleep.

Standard medicine has an increasing array of strong and effective blood pressure lowering drugs. Is this good? I'm not so sure. Certainly there are some very severe hypertensive patients who would be in immediate jeopardy without drug control but a very large number of patients are being treated for "white coat syndrome." Does your blood pressure go up when you are behind a slow driver, or have an argument with someone or when you have to pay a bill that is clearly excessive? Well, your blood pressure may well be up just going to see the physician. In these days of limited time in hospitals there is also limited time in HMO facilities. To give a drug just because your blood pressure is up at the time of examination is a "quick fix". If a betablocker drug is prescribed, your blood pressure will be better in the physician's office but you may well not have the energy or sexual satisfaction that was present before. If a diuretic is prescribed this is usually given in the morning. This is the only time of the day when there is generally no fluid retained. If a diuretic is needed it should be given only once a day and as late as possible so that any accumulated excess fluid is able to be reduced before going to bed. If the diuretic is given too close to bed then you will have excessive urination at night.

Chronic use of diuretics can lead to an escalation of the need for more antihypertensive drugs due to intracellular dehydration. The use of less and less salt in our diet can lead to an actual overall mineral imbalance when coupled with the use of diuretics. I'm not recommending a sudden large increase in table salt usage but a definite increase in water intake plus a balanced salt such as found in miso. This is a form of soy protein which has been fermented and contains Na, Mg, K, I, etc.

When the intracellular mineral and water balance is corrected many hypertensive patients are no longer in need of any drugs.

Neurological Diseases

Stroke

The sudden loss of brain control over an area of the body is loosely called a stroke.

Weakness or paralysis in an extremity can be short lived TIA (Transient Ischemic Attack) or more lasting CVA (Cerebral Vascular Accident).

The patient is taken to the hospital rapidly and after a lot of testing is done you are told what you already know--that the patient had a stroke. The patient may or may not be given a clot buster drug which usually doesn't work. After being given a number of medicines, including coumadin, you are sent to physical therapy for rehabilitation.

It is unfortunate that hospitals do not generally have Hyberbaric Oxygen Therapy available (HBO). Hyperbaric Oxygen Therapy is a medical treatment in which the entire body is exposed to 100% oxygen under approximately double the 14.7 pounds per square inch (sea level air pressure). The oxygen is forced into the blood plasma and is delivered into the cells whether or not a RBC (Red Blood Cell) is present. This rapid return of oxygen to the injured brain cells helps prevent irreversible brain cell loss. We also use special IV's that support the cell energy and antioxidant system in these patients during HBO therapy.

Multiple Sclerosis

The chronic neurological disease multiple sclerosis is usually one in which the patient experiences strange fluctuating numbness, muscle weakness, uncoordinated walking, visual disturbances, etc. These symptoms can vary considerably and this fluctuating character of the symptoms is typical for multiple sclerosis.

As far as we are concerned the causes are poor diet, heavy metal poisoning or hypersensitivity to a metal (usually mercury), and the loss of part of the antioxidant system.

After treating the basic causes, HBO therapy is very useful and is given over a two week period (20 treatments of one hour each).

Cerebral Vascular Disease is just the vascular disease involving the brain's blood supply. If a CVA is involved after the acute therapy is done we generally recommend EDTA chelation therapy (see Heart Disease). The long term use can be very helpful in restoring function of control over limbs. The efficacy of this therapy in CVA recovery is about 65%.

All therapies must start with the maintenance rules.

Alzheimer's Disease

Alzheimer's Disease, when first defined, was described as a rapid loss of mental sharpness in the age group of late 50's and early 60's, leading to severe mental deficiencies. This diagnosis has unfortunately been extended to just about anyone who is not as sharp mentally as before. Many of these changes are caused by Cerebral Artery Disease and the gradual loss of brain cells due to decreased circulation. The now accepted diagnosis of Alzheimer's Disease is described as the neurofibrillary tangles which occur in the brain cells. Mercury vapor inhibits tubulin polymerization into microtubules. In March 2001, mercury was identified as the cause of these anatomic changes in the brain cells of Alzheimer's Disease patients.

Pulmonary Disease

Most people take breathing for granted and only when experiencing shortness of breath, chronic cough, or a productive cough do they take notice.

Shortness of breath can be caused by anything that interferes with the exchange of oxygen and carbon dioxide at the air sacs to the blood supply. Heart failure is the most common cause of shortness of breath.

The chronic lung changes from smoking (emphysema) produce many irreversible lung and air sac changes. Shortness of breath only develops with long term usage. If you smoke, you must stop!

The lungs clean themselves by producing mucous which the person coughs up. This is normal and only when excess is produced does it become a problem. Anything that is foreign that we breathe in hopefully can be be removed by coughing up the mucous. There are muscles in the bronchioles which can contract excessively when stimulated by an irritant producing shortness of breath in a form called asthma.

What is interesting is that just about all of the pulmonary problems can be solved by following the standard maintenance rules. Most drugs can be stopped (in the long run) for most conditions.

Chronic antibiotic use is a bad idea. Most of the bronchitises do not need antibiotics. These should be retained only for serious lung infections caused by bacteria. A simple white blood count and differential can usually identify the acute lung conditions using antibiotics. Antibiotics should not be used for viral diseases. The chronic use of bronchodilators can usually be discontinued when the body is de-stressed and de-poisoned.

There are some chronic conditions of decreased pulmonary function with severe shortness of breath which can be relieved with intravenous use of very dilute hydrogen peroxide. This therapy for the right person allows increased pulmonary function by ejecting small mucous plugs from the small bronchioles. When there are chronic poisonings EDTA chelation therapy may be very useful.

We are designed to have a certain amount of residual air in our lungs at all times. Should you choke on some food, if that air were not there you would not be able to cough up that piece of food. Periodically that dead space residual air needs exchanging and we make a sigh. That sigh helps new air come into those air sacs; however, it involves our taking as deep a breath as possible. We can tend to breathe too deeply if we are not careful. Should you take a deep breath to the extent of your ability and do not for any reason feel satisfied this can produce immediate alarm. The natural inclination is to breathe as deeply as you can after that experience. That is a mistake. Concentrate on getting rid of the air and breathe in only 1/2 way to the maximum. Leave maximum breath for the sigh.

Miscellaneous Tips

Clearing Your Ears

When you are in an airplane, high elevator or in a car coming down from altitude, it is very distressing not to be able to clear your Eustachian tubes. These tubes equalize the air pressure on the ear drums with the outside air pressure. The valves are flutter valves which easily open when you are going up in altitude. There is a muscle in the back of the throat just below the opening to the Eustachian tube which when you yawn usually will clear the problem. You have already found out what I have told you so far but if it doesn't clear then you have a problem. Next maneuver that usually works is to open your mouth and make a good seal with one hand. With your other hand, block your nostrils, then yawn and increase the pressure gradually until the ear "squeaks" with the air coming in. If only one ear is a problem, or after one ear clears, have someone else block the good ear with a finger so the good ear doesn't become overinflated. Good luck. It usually works with a little practice.

Skin

Skin is designed to be dry all of the time and protects itself from moisture by secreting skin oils. There are no skin oil glands on the finger tips; hence, when you are in the water too long, the finger tips look "funny" when the skin becomes waterlogged. Mucous membranes are designed to be moist all of the time and secrete a mucous protection in order to protect from the chronic moisture. The vaginal canal secretes an acidic discharge which protects it from bacterial overgrowth. Soap which is alkaline should be kept away from the vaginal and or rectal areas. When showering, flowing water is all that is needed to clean the surface. Soap in this area causes waterlogging of the mucosa, itching and increases the chance of skin and urinary disease. See, we are better constructed than you thought.

Swallowing Pills

If the capsule or tablet is to be taken without food then place the item under the tongue. Drink some water and the tablet can easily be flipped into the water stream by the tongue. If a supplement is taken with food then after chewing a small amount of food you are ready for swallowing. Place the pill in the food and it is easily swallowed. Always take one pill at a time with food.

The Urinary Tract

If your body has enough water most of the problems of the urinary tract can be prevented and frequently solved. The flow of water keeps the urinary breakdown chemicals of metabolism and excreted toxins from being too concentrated. The production of kidney stones should not occur if there is enough water for the system.

Many people have inadequate thirst and therefore you must evaluate your water status and possibly greatly increase your water intake between meals. One rule of thumb for adults: you may be proud of the fact that you do not have to get up to urinate at night. If you do not get up you are probably not drinking enough water. If the cells in the body are well hydrated, the kidneys will not have antidiuretic hormone from the posterior portion of the pituitary delivered to the kidneys for emergency retention of water. It is interesting that if you urinate only a small amount frequently you can reduce the frequency of urination by drinking more water. While this seems illogical, concentrated urine can be irritating to the bladder and with the increased intake of water the diluted urine will be passed in larger amounts at times of less frequent urination. The major exception to this rule is if there is an outflow blockage producing a reasonable amount of urine which remains in the bladder. This residual urine produces the "small bladder syndrome." Chronic urinary infections in the woman can produce a small urethral opening. This can be corrected by dilation and increasing her immunity. The acidic discharge from the vaginal canal keeps bacterial growth down and the excessive use of alkaline soaps promotes urinary infections. The males outflow blockage usually involves the prostate gland. Benign prostatic hypertrophy pressures the urethra and may produce the "small bladder syndrome." Supplements containing Saw Palmetto and Pygeum Africanis can frequently control this.

About twenty minute before going to bed urinate whether you need to or not and then once again before bed. Since the bladder may not empty itself completely at bedtime, this extra urination frequently will eliminate an extra trip in the middle of the night. If you are urinating too often at night perhaps you are not drinking enough fluids during the day are are retaining salt. If your ankles swell a small amount during the day, put your legs up for a short time (one minute) every couple of hours since vein valves which return the blood may be insufficient.

If you are an adult over about 35 years of age and don't get up to urinate at all at night you are probably not drinking enough water.

Poisoning

Just as you protect yourself from being injured while walking across a busy road or driving your car defensively you must think of protecting yourself from poisonings.

From a chemical and toxic element standpoint we need to realize just how toxic our world is and put protection into the equation in an attempt at longevity. The EPA had created this list of 275 priority hazardous substances. This list is not a list of most toxic substances but a list of the most hazardous (combination of toxicity and the likeliness of your exposure).

While some of these substances are toxic by other mechanisms, the majority affect us by four main mechanisms:

1. Injuring cell wall membranes. Example: radioactive elements and excess iron catalyzing free radical damage (overpowering vitamin E and selenium).

2. Reducing energy production at mitochondria. Example: arsenic poisoning.

3. Poisoning of antioxidant system. Example: cadmium replacing and inactivating zinc in S.O.D.

4. Important component poisoning. Example: cyanide or carbon monoxide poisoning the hemoglobin in the red blood cells.

Each of the 275 poisons on the list has a government-mandated limit before it is considered a toxic level for that particular poison. Unfortunately most of these poisons act on the same place in the cells' metabolism. Despite a different name for each of these 275 poisons the singularity of actions makes them additive in toxicity. Following that same thinking you can not have daily shots of bourbon, scotch, rum, vodka, gin, cordial etc. just because they all have different names. Despite the yearly increase in incidences of major poisonings reported in newspapers and other media, the problem of poisoning is usually disregarded except for alcohol, smoking, and illegal drug addction.

It is interesting that although this list is basically workplace poisons, carbon monoxide is not in the list. It is a major poison. Smoking increases Carbon Monoxide. True acute Carbon Monoxide poisoning is rare in the workplace. This poisoning of the system is not obvious in the chronic lower level state.

Hair Metals: Hair analysis is the most accurate screening method for toxic as well as nutritional minerals. The EPA recommends this for both arsenic and mercury testing. After the use of the chelator DMSA or EDTA urine mercury and other toxic metals can be evaluated with reasonable accuracy.

Whole blood metals can be determined to accuracy of parts per billion. Unfortunately many poisons are trapped in the tissues and may not be found in blood.

The Basic Lymphocyte Profile can identify the immunity system that handles cancer. Natural killer cells and toxic lymphocytes can be followed and increase when the immunity improves with removal of poisons and adding immune stimulants.

Essential fatty acids are the precursors to the regulatory proteins called prostaglandins. These are vital substances in both men and women. Unfortunately they were first found in the prostate gland and hence the name.

Transketolase can give us an excellent evaluation and direction if the person is malnourished. We use a "James Bond" type of test for evaluating the stomach and pancreatic enzymes. The patient swallows a small capsule which is actually an FM radio transmitter. The information is returned by analysis of the radio signal.

Hair, water, urine and blood can be tested with accuracy.

APPENDIX

The following 1999 CERLA List of Priority Hazardous Substances has been obtained from the Agency for Toxic Substances and Disease Registry.

ATSDR Information Center
Division of Toxicology
Mail Stop E-29
1600 Clifton Rd. N.E.
Atlanta, GA 30333
e-mail: ATSDRIC@cdc.gov
Tel. 1-888-422-8737

The 1999 Comprehensive Environmental Response, Compensation, and Liability Act List of Priority Hazardous Substances

RANK	SUBSTANCE NAME
1	ARSENIC
2	LEAD
3	MERCURY
4	VINYL CHLORIDE
5	BENZENE
6	POLYCHLORINATED BIPHENYLS
7	CADMIUM
8	BENZO(A)PYRENE
9	POLYCYCLIC AROMATIC HYDROCARBONS
10	BENZO(B)FLUORANTHENE
11	CHLOROFORM
12	DDT, P,P'-
13	AROCLOR 1260
14	AROCLOR 1254
15	TRICHLOROETHYLENE
16	CHROMIUM, HEXAVALENT
17	DIBENZO(A,H)ANTHRACENE
18	DIELDRIN
19	HEXACHLOROBUTADIENE
20	DDE, P,P'-
21	CREOSOTE

22	CHLORDANE
23	BENZIDINE
24	ALDRIN
25	AROCLOR 1248
26	CYANIDE
27	DDD, P,P'-
28	AROCLOR 1242
29	PHOSPHORUS, WHITE
30	HEPTACHLOR
31	TETRACHLOROETHYLENE
32	TOXAPHENE
33	HEXACHLOROCYCLOHEXANE, GAMMA-
34	HEXACHLOROCYCLOHEXANE, BETA
35	BENZO(A)ANTHRACENE
36	1,2-DIBROMOETHANE
37	DISULFOTON
38	ENDRIN
39	BERYLLIUM
40	HEXACHLOROCYCLOHEXANE, DELTA-
41	AROCLOR 1221
42	DI-N-BUTYL PHTHALATE
43	1,2-DIBROMO-3-CHLOROPROPANE
44	PENTACHLOROPHENOL
45	AROCLOR 1016
46	CARBON TETRACHLORIDE
47	HEPTACHLOR EPOXIDE
48	XYLENES, TOTAL
49	COBALT
50	ENDOSULFAN SULFATE
51	DDT, O,P'-
52	NICKEL
53	3,3'-DICHLOROBENZIDINE
54	DIBROMOCHLOROPROPANE
55	ENDOSULFAN, ALPHA
56	ENDOSULFAN
57	BENZO(K)FLUORANTHENE
58	AROCLOR
59	ENDRIN KETONE
60	CIS-CHLORDANE
61	2-HEXANONE
62	TOLUENE
63	AROCLOR 1232

64	ENDOSULFAN, BETA
65	METHANE
66	TRANS-CHLORDANE
67	2,3,7,8-TETRACHLORODIBENZO-P-DIOXIN
68	BENZOFLUORANTHENE
69	ENDRIN ALDEHYDE
70	ZINC
71	DIMETHYLARSINIC ACID
72	DI(2-ETHYLHEXYL)PHTHALATE
73	CHROMIUM, TRIVALENT
74	METHYLENE CHLORIDE
75	NAPHTHALENE
76	METHOXYCHLOR
77	1,1-DICHLOROETHENE
78	AROCLOR 1240
79	BIS(2-CHLOROETHYL)ETHER
80	1,2-DICHLOROETHANE
81	2,4-DINITROPHENOL
82	2,4,6-TRINITROTOLUENE
83	2,4,6-TRICHLOROPHENOL
84	CHLORINE
85	CYCLOTRIMETHYLENETRINITRAMINE (RDX)
86	1,1,1-TRICHLOROETHANE
87	ETHYLBENZENE
88	1,1,2,2-TETRACHLOROETHANE
89	THIOCYANATE
90	ASBESTOS
91	4,6-DINITRO-O-CRESOL
92	URANIUM
93	RADIUM
94	RADIUM-226
95	HEXACHLOROBENZENE
96	ETHION
97	THORIUM
98	CHLOROBENZENE
99	BARIUM
100	2,4-DINITROTOLUENE
101	FLUORANTHENE
102	RADON
103	RADIUM-228
104	THORIUM-230
105	DIAZINON

106	BROMINE
107	1,3,5-TRINITROBENZENE
108	URANIUM-235
109	TRITIUM
110	URANIUM-234
111	THROIUM-228
112	N-NITROSODI-N-PROPYLAMINE
113	CESIUM 137
114	HEXACHLOROCYCLOHEXANE, ALPHA-
115	CHRYSENE
116	RADON-222
117	POLONIUM-210
118	CHRYSOTILE ASBESTOS
119	THORIUM-227
120	POTASSIUM-40
121	COAL TARS
122	PLUTONIUM-238
123	THORON (RADON-220)
124	COPPER
125	STRONTIUM-90
126	COBALT-60
127	METHYLMERCURY
128	CHLORPYRIFOS
129	LEAD-210
130	PLUTONIUM-239
131	PLUTONIUM
132	AMERICIUM-241
133	IODINE-131
134	AMOSITE ASBESTOS
135	GUTHION
136	BISMUTH-214
137	LEAD-214
138	CHLORDECONE
138	PLUTONIUM-240
138	TRIBUTYLTIN
141	MANGANESE
142	S,S,S-TRIBUTYL PHOSPHOROTRITHIOATE
143	SELENIUM
144	POLYBROMINATED BIPHENYLS
145	DICOFOL
146	PARATHION
147	HEXACHLOROCYCLOHEXANE, TECHNICAL

148 PENTACHLOROBENZENE
149 TRICHLOROFLUOROETHANE
150 TREFLAN (TRIFLURALIN)
151 4,4’-METHYLENEBIS(2-CHLOROANILINE)
152 1,1-DICHLOROETHANE
153 DDD,O,P’-
154 HEXACHLORODIBENZO-P-DIOXIN
155 HEPTACHLORODIMENZO-P-DIOXIN
156 2-METHYLNAPHTHALENE
157 1,1,2-TRICHLOROETHANE
158 AMMONIA
159 ACENAPHTHENE
160 1,2,3,4,6,7,8,9-OCTACHLORODIBENZOFURAN
161 PHENOL
162 TRICHLOROETHANE
163 CHROMIUM(VI) TRIOXIDE
164 1,2-DICHLOROETHENE, TRANS-
165 HEPTACHLORODIBENZOFURAN
166 HEXACHLOROCYCLOPENTADIENE
167 1,4-DICHLOROBENZENE
168 1,2-DIPHENYLHYDRAZINE
169 CRESOL, PARA-
170 1,2-DICHLOROBENZENE
171 LEAD-212
172 OXYCHLORDANE
173 2,3,4,7,8-PENTACHLORODIBENZOFURAN
174 RADIUM-224
175 ACETONE
176 HEXACHLORODIBENZOFURAN
177 BENZOPYRENE
177 BISMUTH-212
179 AMERICIUM
179 CESIUM-134
179 CHROMIUM-51
182 TETRACHLOROPHENOL
183 CARBON DISULFIDE
184 CHLOROETHANE
185 INDENO(1,2,3-CD)PYRENE
186 DIBENZOFURAN
187 P-XYLENE
188 2,4-DIMETHYLPHENOL
189 AROCLOR 1268

190	1,2,3-TRICHLOROBENZENE
191	PENTACHLORODIBENZOFURAN
192	HYDROGEN SULFIDE
193	ALUMINUM
194	TETRACHLOROETHANE
195	CRESOL, ORTHO-
196	1,2,4-TRICHLOROBENZENE
197	HEXACHLOROETHANE
198	BUTYL BENZYL PHTHALATE
199	CHLOROMETHANE
200	VANADIUM
201	1,3-DICHLOROBENZENE
202	TETRACHLORODIBENZO-P-DIOXIN
203	2-BUTANONE
204	N-NITROSODIPHENYLAMINE
205	PENTACHLORODIBENZO-P-DIOXIN
206	2,3,7,8-TETRACHLORODIBENZOFURAN
207	SILVER
208	2,4-DICHLOROPHENOL
209	1,2-DICHLOROETHYLENE
210	BROMOFORM
211	ACROLEIN
212	CHROMIC ACID
213	2,4,5-TRICHLOROPHENOL
214	NONACHLOR, TRANS-
215	COAL TAR PITCH
216	PHENANTHRENE
217	NITRATE
218	ARSENIC TRIOXIDE
219	NONACHLOR, CIS-
220	HYDRAZINE
221	TECHNETIUM-99
222	NITRITE
223	ARSENIC ACID
224	PHORATE
225	BROMODICHLOROETHANE
225	DIMETHOATE
227	STROBANE
228	NALED
229	ARSINE
230	4-AMINOBIPHENYL

230 PYRETHRUM
230 TETRACHLOROBIPHENYL
233 DIBENZOFURANS, CHLORINATED
233 ETHOPROP
233 NITROGEN DIOXIDE
236 CARBOPHENOTHION
236 THORIUM-234
238 DICHLORVOS
238 OZONE
238 PALLADIUM
241 CALCIUM ARSENATE
241 CARBON-14
241 EUROPIUM-154
241 KRYPTON-85
241 MERCURIC CHLORIDE
241 SODIUM-22
241 STRONTIUM-89
241 SULFUR-35
241 URANIUM-233
250 2,4-D ACID
251 ANTIMONY
252 CRESOLS
253 PYRENE
254 2-CHLOROPHENOL
255 DICHLOROBENZENE
256 FORMALDEHYDE
257 N-NITROSODIMETHYLAMINE
258 CHLORODIBROMOMETHANE
259 SUTAN
260 DICHLOROETHANE
261 1,3-DINITROBENZENE
262 DIMETHYL FORMAMIDE
263 1,3-DICHLOROPROPENE, CIS-
264 ETHYL ETHER
265 4-NITROPHENOL
266 1,3-DICHLOROPROPENE, TRANS-
267 TRICHLOROBENZENE
268 FLUORIDE
269 1,2-DICHLOROPROPANE
270 2,6-DINITROTOLUENE
271 METHYL PARATHION

272	METHYL ISOBUTYL KETONE
273	OCTACHLORODIBENZO-P-DIOXIN
274	STYRENE
275	FLUORENE

Substances were assigned the same rank when two (or more) substances received equivalent total scores.

ARSENIC

#1 on the EPA's list of Most Hazardous Substances in 1999

Arsenic is a natually occurring element that is widely distributed in the crust of the earth. Exposure to arsenic is mostly likely to occur in the workplace, near hazardous waste sites, or in areas with high levels naturally. In the environment, arsenic combines with oxygen, chlorine and sulfur to form inorganic arsenic compounds. Inorganic compounds are used primarily to preserve wood. Organic arsenic compounds are used as pesticides, especially cotton plants.

Arsenic cannot be destroyed in the environment, it only changes form. Arsenic in the air will settle to the ground or is washed out of the air by rain. Many arsenic compounds will dissolve in water. Consequently, arsenic in the water supply is of great concern. Fish and shellfish can accumulate arsenic, although in a form that may not be as toxic as other sources.

People are exposed to arsenic through food, water and air. Of particular concern is air in certain workplace settings. Wood treated with arsenic is used commonly as building material, particularly in the type of wood used in decks. Breathing sawdust or burning smoke from such wood exposes a person to arsenic. If such wood is used in home projects, it is important to wear dust masks, gloves and protective clothing. People who live near uncontrolled hazardous waste sites containing arsenic are at risk as are people who live in areas with unusually high natural levels of arsenic in rock. To reduce risk, limit contact with the soil and obtain pure, fresh water from outside sources.

Breathing high levels of inorganic arsenic can cause throat and lung irritation. Ingesting high levels can result in death. Lower levels of exposure can cause nausea, vomiting, decreased blood cell production, abnormal heath rhythm, blood vessel damage and tingling in the hands and feet. Prolonged exposure can result in darkening of the skin and the appearance of small wart-like growths on the body. Several studies have shown that inorganic arsenic can increase cancer risk.

It is possible to test for arsenic levels in the body through blood, urine, hair or fingernails.

LEAD

#2 on the EPA's list of Most Hazardous Substances in 1999

Lead is a metal that occurs naturally in small amounts in the crust of the earth. Although it can be found in all parts of our environment, much of it occurs as a result of human activities such as burning fossil fuels, mining and manufacturing. Lead is used in the production of batteries, ammunition, metal products such as solder and pipes, and devices to shield X-rays. Due to health concerns, lead from gasoline, paints, ceramic products, caulking and pipe solder has been reduced substantially in recent years.

Lead itself does not break down, but lead compounds can be changed by sunlight, air and water. Lead released into the air may travel great distances in the air before settling to the ground. Once lead falls onto soil, it sticks to soil particles. Depending on the lead compound and oil characteristics, lead may find its way into groundwater.

Lead exposure commonly occurs as a result of breathing contaminated workplace air or dust, eating contaminated foods or drinking contaminated water. Children can be exposed to lead from swallowing chips or dust of lead-based paint that peels off the walls of older homes built before 1978, chewing on objects painted with lead-based paint or playing outdoors in soil that has been contaminated. Children should have their hands and faces washed often to remove lead dusts and soil and homes should be cleaned regularly. Soil contamination can occur from naturally occurring lead, lead that has settled into the soil from auto emissions or from paint. Some healthcare products or folk remedies can contain lead as can some types of paints and pigment that are used as make-up or hair coloring. Hobbies using lead, such as stained glass work, expose one to the substance.

Lead can affect almost every organ and system in the body. The central nervous system, particularly in children, is most sensitive. The effects are the same whether breathed or swallowed. High lead levels may cause decreased reaction time, cause weakness in the fingers, wrists or ankles, and affect the memory. It may cause anemia, damage the reproductive system and kidneys. Certain lead compounds are known carcinogens. Children are more vulnerable to lead poisoning than adults. Signs of lead toxicity in a child can include blood anemia, severe stomach ache, muscle weakness and brain damage. Even at low levels of exposure, lead can affect a child's mental and physical growth. Unborn children can be exposed to lead through their mothers. Effects can include premature births, small body size, decreased mental ability, learning difficulties and reduced growth.

Lead levels in the body can be tested using blood, hair or urine.

MERCURY

#3 on the EPA's list of Most Hazardous Substances in 1999

Mercury is a naturally occurring metal which has several forms. Metallic mercury is a shiny, silver-white, odorless liquid. If heated, it becomes a colorless, odorless gas. Mercury combines with other elements to form inorganic mercury compounds or "salts." It also combines with carbon to make organic mercury compounds. The most common of these, methylmercury, is produced mainly by bacteria in the water and soil. Inorganic mercury enters the air from mining ore deposits, burning coal and waste, and from manufacturing plants. It can enter the water or soil from natural deposits, disposal of wastes, and volcanic activity. Methylmercury builds up in the tissues of fish; the larger and older the fish the higher the mercury levels.

Sources of mercury include mercury amalgam dental fillings, eating fish or shellfish contaminated with mercury, breathing vapors from spills, incinerators and industries that burn mercury-containing fuels, breathing contaminated workplace air or having skin contact during use in the workplace.

The nervous system is highly sensitive to all forms of mercury. Methylmercury and metallic mercury vapors are more harmful than other forms because more mercury in these forms reaches the brain. Exposure to high levels of metallic, inorganic, or organic mercury can permanently damage the brain, kidneys and developing fetus. Effects on brain function may result in irritability, shyness, tremors, changes in vision or hearing, and memory problems. Short-term exposure to high levels of metallic mercury vapors may cause effects including lung damage, nausea, vomiting, diarrhea, increases in blood pressure or heart rate, skin rashes and eye irritation. The EPA has determined that mercuric chloride and methylmercury are possible human carcinogens. Very young children are more sensitive to mercury than adults. Mercury can be passed to the fetus from the mother and to nursing children through breast milk. Harmful effects from these sources of mercury can include brain damage, mental retardation, poor coordination, blindness, seizures and inability to speak.

People with mercury amalgam dental fillings should have their fillings replaced by a dentist familiar with toxic-free dentistry. Spilled mercury should never be vacuumed to prevent vaporization. If a large amount of mercury has been spilled, it is essential to contact the health department. Children should be warned not to play with shiny, silver liquids. Older medicines that contain mercury should be properly disposed of. Be aware of wildlife and fish advisories from public health or natural resources departments.

Tests using blood, hair, or urine samples are available to test body mercury levels.

VINYL CHLORIDE

#4 on the EPA's list of Most Hazardous Substances in 1999

Vinyl chloride is a colorless, flammable gas at normal temperatures with a mild, sweet odor. It is a manufactured substance that is used to make polyvinyl chloride (PVC). PVC is used to make a variety of plastic products, including pipes, wire and cable coatings, furniture and automobile upholstery. Vinyl chloride also results from the breakdown of other substances. Liquid vinyl chloride evaporates easily into the air. When in the air, it can break down into other substances, some of which are harmful. Small amounts of vinyl chloride can dissolve in water. Vinyl chloride formed from the breakdown of other chemicals can enter groundwater. It is unlikely to build up in plants or animals.

The most common means of exposure are breathing vinyl chloride that has been released from plastics industries, hazardous waste sites and landfills, breathing it in the air or during contact with skin or eyes in the workplace and drinking water from contaminated wells.

Breathing high levels of vinyl chloride can cause dizziness or sleepiness with the potential of unconsciousness, and, in the case of extremely high levels, death. Most of the studies on long term exposure are about workers that make or use the substance. People who work with vinyl chlorides have developed nerve damage and immune reactions. Other workers have developed problems with the blood flow in their hands. In some cases, the bones in the tips of the fingers have broken down. The effects of drinking high levels of vinyl chloride are unknown. If spilled on the skin, it causes numbness, redness and blisters. Long term animal studies link exposure to birth defects, stillbirths and miscarriages. The Department of Health and Human Services has determined that vinyl chloride is a known human carcinogen.

The results of breath testing can sometimes show exposure to vinyl chloride. Better information is obtained by measuring a breakdown product of vinyl chloride in the urine shortly after exposure. Testing of genetic material can indicate vinyl chloride exposure.

BENZENE

#5 on the EPA's list of Most Hazardous Substances in 1999

Benzene is a colorless liquid with a sweet odor. It evaporates into the air very quickly and dissolves slightly in water. It is highly flammable and is formed from both natural processes and human activities. Benzene is widely used in the United States, ranking in the chemicals which are used to make plastics, resins, nylon and synthetic fibers. Benzene is also used to make some types of rubbers, lubricants, dyes, detergents, drugs, and pesticides. Natural sources of benzene include volcanoes and forest fires. Benzene is also a natural part of crude oil, gasoline and tobacco smoke, which is a major source.

Industrial processes are the main sources of benzene in the environment and it can pass from the air into the water and soil. Benzene in the air can also attach to rain or snow and be carried to the ground. It breaks down slowly in the ground and can pass into the underground water. Benzene does not build up in plants or animals. Outdoor air conatins low levels of benzene from tobacco smoke, automobile service stations, motor vehicle exhaust and industrial emissions. Indoor air usually contains higher levels of benzene from glues, paints, furniture wax and detergents. Air around hazardous waste sites or gas stations will contain higher levels. Well water can also become contaminated due to leakage from underground storage tanks or hazardous waste sites. People working in industries that make or use benzene may be exposed to the highest levels.

Exposure to high levels of benzene can cause drowsiness, dizziness, rapid heart rate, headaches, tremors, confusion and unconsciousness. Very high levels can cause death. Long term exposure can damage the bone marrow and affect red blood cells leading to anemia. It can also cause excessive bleeding and affect the immune system. Animal studies have shown low birth weights, delayed bone formation and bone marrow damage in pregnancy. The Department of Health and Human Services has determined that benzene is a known human carcinogen.

Breath tests can measure benzene levels shortly ater exposure. Blood tests will indicate recent exposure. Urine testing may be performed for benzene metabolites, but must be performed soon after exposure.

POLYCHLORINATED BYPHENYLS (PCB)

#6 on the EPA's list of Most Hazardous Substances in 1999

Polychlorinated biphenyls, PCBs, are mixtures of chlorinated compounds. They can exist as a vapor in the air or be a liquid or solid. They have no smell or taste. Many commercial PCB mixtures are known in the United States by the trade name Aroclor. PCBs have been used as coolants and lubricants in transformers and electrical equipment. The manufacture of PCBs was stopped in this country in 1977 because of evidence of buildup in the environment. Products made before 1977 that may contain PCBs include old fluorescent lighting fixtures, electrical devices containing PCB capacitors and old microscope and hydraulic oils.

PCBs enter the air, water and soil during their manufacture, use and disposal; accidental spills and leaks during transport; and from leaks or fires in products containing PCBs. They can be released into the environment from waste sites or improper disposal methods. PCBs do not readily break down in the environment and can travel long distances in the air. They bind with soil and also concentrate in bottom sediment in water sources. PCBs are taken up by small organisms and fish in water and other animals that eat these aquatic animals as food accumulating in fish and marine mammals in levels much higher than found in the water. People may be exposed from eating fish or wildlife caught from contaminated locations. Children should not play with old appliances, electrical equipment or transformers and avoid dirt near hazardous waste sites or near where there was a transformer fire. Workplace exposure can contaminate clothing or tools.

People exposed to large amounts of PCBs can develop skin conditions such as acne and rashes. There may also be liver damage. Effects of PCBs in animals include changes in the immunce system, behavioral alterations and impaired reproduction. The Department of Health and Human Services, the EPA and the international Agency for Research on Cancer have all concluded that PCBs may be carcinogens.

Tests exist to measure levels of PCBs but are not routinely conducted. Most people normally have low levels of PCBs in the body from the environmental exposure.

CADMIUM

#7 on the EPA's list of Most Hazardous Sybstances in 1999

Cadmium is a natural element in the earth's crust. It is usually found as a mineral combined with other elements. All soil and rocks, including coal and mineral fertilizers, contain some cadmium. Most cadmium used in the United States is extracted during the production of other metals such as zinc, lead and copper. Cadmium does not corrode easily and has many uses, including batteries, pigments, metal coatings and plastics.

Cadmium enters the air from mining, industry and burning coal and household wastes. Particles in the air can travel long distances before falling to the ground or water. It can also enter water and soil from disposal and spills or leaks at hazardous waste sites. It binds strongly to soil particles. Some cadmium dissolves in water. It does not break down in the environment but can change forms. Fish, plants and animals take up cadmium from the environment. It stays in the body a very long time and can build up from many years of exposure to low levels. Exposures can occur from breathing contaminated workplace air, especially due to battery manufacturing, metal soldering or welding. All foods contain low levels of cadmium but levels are highest in shellfish, liver and kidney. Breathing cadmium in cigarette smoke doubles the average daily intake. Exposures can occur from drinking contaminated water. Also, a person can be exposed by breathing contaminated air near the burning of fossil fuels or municipal waste.

Breathing high levels of cadmium severely damages the lungs and can cause death. Eating food or drinking water with very high levels can result in severe stomach irritation, vomiting, and diarrhea. Long term exposure to lower levels leads to a buildup of cadmium in the kidneys resulting in kidney disease, lung damage and fragile bones. Cadmium may affect birth weight, skeletal development, behavior and learning ability. Animal studies show that more cadmium is absorbed into the body if the body is low in calcium, protein or iron or is high in fat. The Depatrment of Health and Human Services has determined that cadmium and the compounds may be carcinogenic.

Tests are available to measure cadmium in blood, urine, hair or nails.

POLYCYCLIC AROMATIC HYDROCARBONS (PAH)

#9 on the EPA's list of Most Hazardous Substances in 1999

Polycyclic aromatic hydrocarbons (PAHs) are a group of over 100 different chemicals that are formed during the incomplete burning of oil, gas, coal, garbage or other organic substances like tobacco or charbroiled meat. PAHs are usually found as a mixture containing two or more of these compounds, such as soot. Some PAHs are manufactured. PAHs are found in coal tar, crude oil, creosote and roofing tar. A few are used in medicines or to make dyes, plastics and pesticides.

PAHs enter the air primarily as releases from volcanoes, forest fires, burning coal and automobile exhaust. They can evaporate into air from soil or surface waters. They can enter water through discharges from industrial and wsatewater treatment plants. Most PAHs do not dissolve easily in water but stick to solid particles and settle to the bottoms of lakes or rivers. PAHs can move through soil to contaminate underground water. PAH contents of plants and animals may be much higher than the levels found in soil or water. Exposures can occur in the workplace of cooking, coal tar and asphalt production plants, smokehouses, and municipal trash incineration facilities. Exposures can also occur from breathing air containing cigarette smoke, wood smoke, vehicle exhausts, asphalt roads or agricultural burn smoke. Grilled or charred meats, contaminated cereals, flour, bread, vegetables, fruits or meats and processed or pickled foods are sources. It is possible to be exposed by coming in contact with air, water or soil near hazardous waste sites or drinking contaminated water or cow's milk. Nursing infants of mothers living near hazardous waste sites may be exposed through mother's milk.

Mice fed PAH during pregnancy had difficulty reproducing as did their offspring. There were also increased incidences of birth defects and lower body weights. Animal studies have also shown that PAHs can cause harmful effects on the skin, body fluids and ability to fight disease. The Department of Health and Human Services has determined that some PAHs may be carcinogenic.

There are special tests to determine evidence of PAH presence in the body but they cannot determine extent or source of exposure.

RECOMMENDED READING

Brodeur, Paul. 1993. *The Great Power-Line Cover-Up.* Little, Brown and Company

Carter, James P., M.D., P.H.D. 1993. *Racketeering in Medicine—The Suppression of Alternatives.* Hampton Roads, Charlottesville, VA

Cranton, Elmer M., M.D. 2001. *Bypassing Bypass Surgery.* Hampton Roads Publishing, Charlottesville, VA

Cranton, Elmer M.D. and Fryer, William. 1996. *Resetting the Clock.* M. Evans and Co. Inc., NY

Duffy, William. 1975. *Sugar Blues.* Warner Books, Inc., NY

Goldbeck, Nikki and David. 1984. *Nikki and David Goldbeck's American Wholefoods Cuisine.* Plume Publishing (Penguin Books), NY

Huggins, Hal, D.D.S., 1993. *It's All in Your Head—The Link Between Mercury Amalgams and Illness.* Avery Publishing

Jerome, Frank J., D.D.S. 2000. *Tooth Truth—A Patient's Guide to Metal-Free Dentistry.* New Century Press, Chula Vista, CA

Lee, John R., M.D. 1999. *What Your Doctor May Not Tell You About Menopause.* Warner Books, NY

Lonsdale, Derrick, M.D. 1994. *Why I Left Orthodox Medicine—Healing for the 21st Century.* Hampton Roads Publishing, Charlottesville, VA

McGee, Charles T., M.D. 1993. *Heart Fraud—The Misapplication of High Technology in Heart Disease.* Medipress, Coeur d'Alene, ID

Neubauer, Richard A., M.D., and Walker, Morton, D.P.M. 1998. *Hyperbaric Oxygen Therapy.* Avery Publishing

Proctor, Robert N. 1995. *Cancer Wars--How Politics Shape What We Know and Don't Know About Cancer.* Basic Books, NY

Robinson, Jeffrey. 2001. *Prescription Games.* Pennstreet Ltd., Toronto

Schlosser, Eric. 2001. *Fast Food Nation.* Houghton Mifflin Company, Boston & New York

Schwartz, George. 1999. *In Bad Taste: The MSG Symptom Complex.* Health Press

Yiamouyiannis, John, Dr. 1986. *Flouride the Aging Factor: How to Recognize and Avoid the Devastating Effects of Fluoride.* Health Action Press

FREE RADICAL PATHOLOGY IN AGE-ASSOCIATED *DISEASES: TREATMENT WITH EDTA CHELATION,* NUTRITION, AND ANTIOXIDANTS

Elmer M. Cranton, MD
James P. Frackelton, MD

ABSTRACT: Recent discoveries in the field of free radical pathology provide a coherent, unifying scientific basis to explain many of the diverse benefits reported from treatment with EDTA chelation therapy. The free radical concept provides a scientific basis for treatment and prevention of the major causes of disability and death, including atherosclerosis, dementia, cancer, arthritis, and numerous other diseases. EDTA chelation therapy, hyperbaric oxygen therapy, applied clinical nutrition, nutritional supplementation, physical exercise, and moderation of health-destroying habits all have common therapeutic mechanisms that reduce free radical causes of many age-associated diseases.

Chelation therapy with intravenous ethylenediaminetetraacetate (EDTA) has been practiced by an increasing number of physicians for over three decades. Published studies describe beneficial results using intravenous EDTA as therapy for patients with chronic degenerative diseases.[1-47] Reports of renal injury and other adverse effects from this therapy are now known to have been caused by doses exceeding 50 mg/Kg/day, by preexisting kidney disease and by heavy metal toxicity.[48-56] No published studies utilizing currently accepted treatment procedures reported negative results. The often cited Kitchell report contained an adverse conclusion,[57] although data from that study was quite favorable to the therapy. A careful review[58] of the data in that report does not support the authors' conclusion. Research with laboratory animals provides further support for the effectiveness of EDTA chelation therapy.[59-65]

FREE RADICAL CAUSES OF DEGENERATIVE DISEASE

Recent discoveries in the field of free radical pathology provide a coherent and scientific basis to explain many of the report benefits resulting

---

Elmer M. Cranton, M.D., is a past Vice-President of the American Academy of Medical Preventics; Past President of the American Holistic Medical Association; Charter Fellow, American Academy of Family Physicians; Diplomate, American Board of Family Practice; and Diplomate, American Board of Chelation Therapy. James P. Frackelton, M.D., is President Elect of the American Academy of Medical Preventics; Instructor in Chelation Therapy, American Board of Chelation Therapy; Diplomate, American Board of Chelation Therapy; and past Chairman, Department of Family Practice, Fairview General Hospital, Cleveland, Ohio.

from EDTA chelation therapy, usually in conjunction with nutritional and life-style changes. The field of free radical biochemistry is as revolutionary and profound in its implications for development of effective treatments for infectious diseases. The free radical concept explains contradictory epidemiologic and clinical observations and provides a scientific rationale for treatment and prevention of many of the major causes of long-term disability and death: atherosclerosis, dementia, cancer, arthritis, and other age-related diseases.[66-75]

Methods for the detection and measurement of free radicals have only recently been developed.[76-78] EDTA chelation therapy and hyperbaric oxygen therapy, when properly administered in a program of physical exercise, applied clinical nutrition, and moderation of health-destroying habits, all have common therapeutic mechanisms that reduce free radical damage.

WHAT ARE FREE RADICALS?

A free radical has an unpaired electron in an outer orbit, causing it to be highly unstable and to react almost instantaneously with any substance in its vicinity.[79-80] The half life of biologically active free radicals is measured in micro-seconds.[68] These reactions often produce a cascade of free radicals in a multiplying effect.[66,69,70,77,79,81,82] Free radicals are very unstable and highly reactive molecules and molecular fragments that react aggressively with other molecules, rapidly creating new compounds. Harmful effects of high-energy ionizing radiation (ultraviolet light, x-rays, gamma rays, cosmic radiation) result because radiation knocks electrons out of orbiting pairs, causing free radicals in living tissues.[78,83-87] Free radicals in cell membranes produce pathologic lipid peroxides, oxyarachidonate and oxycholesterol products.[66,87-90] Protection against pathologic free radicals is provided by dietary and endogenous antioxidants.[66,67,69,70,71,73,74,89,91,92]

Many free radical chemical reactions occur normally in the body and are necessary for health.[66-71,73-75,93-96] This process might be thought of as a form of controlled "internal radiation." Highly reactive free radicals that are continuously produced within human cells include hydroxyl radicals, superoxide radicals, and excited or singlet state oxygen.[67-75] Free radicals react to produce hydrogen peroxide, lipid peroxides, and other peroxides. Peroxides are metastable, highly reactive molecules that react rapidly, producing additional organic free radicals in biologic tissues.[79,80]

To prevent uncontrolled multiplication of free radicals, human cells utilize more than a dozen antioxidant control systems that regulate necessary and desirable free radical reactions.[66-74,77,78,84,89,91,97-102] Control mechanisms involve several enzymes, including catalase, superoxide dismutase, and glutathione peroxidase, in conjunction with vitamins C and E, beta carotene, the trace element selenium, and others. When functioning properly, these antioxidant systems suppress excessive free radical reactions, allowing desirable biological effects without unwanted cellular and molecular damage. Without these control systems, free radicals multiply rapidly, much like a nuclear chain reaction, disrupting cell membranes, damaging enzymes, interfering with both active and passive transport across cell membranes and causing mutagenic damage to nuclear material. An abnormally functioning or malignant cell may result.[66,69,70,77,81,82,103,104]

The activity of the free radical control enzyme superoxide dismutase (SOD) in various mammalian species is directly proportional to their life spans. Humans have the highest activity of SOD. It is the fifth most prevalent protein in humans.[69-70] Life expectancy therefore seems highly dependent on effective free radical regulation.

There are a number of nonenzymatic free radical scavengers, some of which can be stoichiometrically consumed on a one-to-one ration when neutralizing free radicals. These include beta carotene (pro-vitamin A), vitamin E, vitamin C, glutathione, cysteine, methionine, tyrosine, cholesterol, glucocorticosteroids, and selenium.

Enzymes involved in free radical protection require trace elements or B vitamins as co-enzymes. The trace elements copper, zinc, and manganese are essential to the superoxide dismutases; selenium is essential to glutathione peroxidase; and iron is necessary for catalase and some forms of peroxidase. Adequate dietary intake of these trace nutrients is necessary for protection against free radical produced diseases.

IDENTIFYING FREE RADICALS

Free radicals are highly unstable. They react rapidly in living tissues and therefore have low steady state concentrations. They rarely reach levels sufficient for direct analysis.[66,70] Newly developed instruments have revealed the importance and extent of free radical damage in tissues. It is now possible to detect the presence of free radicals using electron paramagnetic resonance spectroscopy (EPR).[77-78] However, free radical effects can be measured more precisely by analyzing the end products of free radical reactions, using gas chromatography, mass spectroscopy, and high-performance liquid chromatography. Cross-linkages, damaged collagen, lipid peroxides, oxyarachidonate, oxidized cholesterol, mucopolysaccharide breakdown products, lipofuscin, ceroid, and increased melanin all result from undesirable free radical reactions and can be readily quantitated.[66,69,70,78,105,106]

By sifting through molecular wreckage left in the wake of free radicals, it is possible to determine the type and extent of ongoing free radical reactions. For example, free radical pathology in the central nervous system (CNS) can be assessed by the rate of disappearance of cholesterol. Cholesterol is not metabolized in the central nervous system. The only way for cholesterol to diminish in the CNS is through oxidation caused by free radicals.[69,70,107,108]

CHOLESTEROL METABOLISM

Cholesterol is an antioxidant and free radical scavenger liberally dispersed in cell walls that protects cell membranes.[78,107] Cholesterol also acts as a precursor to the many steroid hormones and vitamin D. Vitamin D is normally produced in the skin by exposure of cholesterol to ultraviolet radiation from sunlight. Ultraviolet light is a form of ionizing radiation that produces free radicals in living tissues.

Total cholesterol (reflected by blood cholesterol) is determined primarily by cholesterol synthesis, not by dietary cholesterol intake.[70] Serum cholesterol levels are now believed to be indicators of free radical damage, and for that reason, correlate with the risk of atherosclerosis.[70] Cholesterol is synthesized in the body as needed, and the need is greater in those at risk. In Western cultures affected with epidemic free radical diseases, blood cholesterol levels increase with age. Subsequent alterations of LDL receptor sites and underlying hereditary factors also affect blood cholesterol levels.

Free radicals peroxidize cholesterol into a variety of degradation products.[66,68,69,70,78,109,110] Oxidized cholesterol is bound selectively to low density lipoproteins, LDL cholesterol, while unoxidized (antioxidant) cholesterol is bound to high density lipoproteins, HDL cholesterol.[69,70] LDL cholesterol is toxic to cells.

Laboratory research at the Cleveland Clinic has demonstrated that both EDTA and the antioxidant glutathione prevent LDL from becoming cytotoxic.[109]

The oxidation of cholesterol results in end products with varying toxicities.[66-70,109-112] Some of these products have vitamin D activity, which may produce localized vitamin D excess in tissues and macrophages.[110] Ectopic calcium may then be deposited in response to the localized excess vitamin D activity. Free radicals also increase tissue calcium by damaging the integrity of cell membranes and by impairing active transport mechanisms.

Dietary restriction of cholesterol and medications to reduce blood cholesterol have been counterproductive in the treatment of atherosclerosis because the antioxidant role of cholesterol has not been recognized. Unoxidized cholesterol is widely dispersed in cell membranes as a protective factor against atherosclerosis, cancer, and other free radical induced diseases. In its antioxidant form, it is not the harmful substance we have been led to believe. Cholesterol is a fat soluble, and restriction of dietary cholesterol also results in a reduction of total dietary fats. This is beneficial not because of cholesterol restriction (unless the cholesterol has been oxidized in food processing, as is often the case), but because of reduced intake of fats containing lipid peroxides, resulting in reduction of free radical pathology.[70]

There is a statistical correlation between very low blood cholesterol levels and increased risk of some types of cancer. If adequate amounts of cholesterol cannot be produced, because of nutritional or metabolic factors, one defense against free radical causes of cancer has been reduced. Free radicals are both primary initiators and subsequent promoters of malignant change. Thus, high fat diets, rich in lipid peroxides, are known to be a major cause of cancer.[66,69,70]

A number of other cholesterol-derived steroids such as glucocorticosteroids, dehydroepiandrosterone (DHEA), and estrogen are also effective free radical scavengers.[78] DHEA and estrogen both diminish rapidly in middle age, a time when the incidence of free radical diseases increases.

ESSENTIAL FREE RADICAL REACTIONS

Human life cannot exist without a balance of carefully regulated free radical reactions. Organisms that live in the presence of oxygen must have free radical protection or they quickly die. Cellular respiration requires transfer of electrons across mitochondrial membranes. A superoxide radical is produced for each electron transferred. Protection against those superoxide radicals is provided by mitochondrial superoxide dismutase (SOD), a manganese-containing enzyme. The average American diet contains suboptimal amounts of manganese.[113] SOD elsewhere in cytoplasm contains zinc and copper, also marginal to deficient in many American diets.[113] If phospholipids in cell membranes are not protected from free radical oxidation, activity of the associated enzymes is lost.[114]

Detoxification of many chemicals, including drugs, artificial colorings and flavorings, petrochemicals, and inhaled fumes, is performed in the endoplasmic reticulum of liver cells and other organs by reactions involving cytochrome P-450 and other enzymes. These detoxification reactions produce both hydroxyl free radicals and peroxides.[66,69,70,93-95] Glutathione peroxidase, vitamin C, and other antioxidants are present to prevent free radical proliferation. Increased exposure to drugs and chemicals causes increased production of free radicals, which can exceed the threshold of antioxidant protection.[66,69,70] Excess free radicals then proliferate in chain reactions, multiplying their damage by a million times or more.[66,70]

Synthesis of prostaglandins and leukotrienes from unsaturated fatty acids also results in release of free radicals,[70,74,96] and in the presence of excessive free radicals, the synthesis of prostaglandins may be imbalanced. For example, production of thromboxane increases, while prostacyclin decreases in the presence of lipid peroxides.[66,69,70]

Leukocytes and macrophages normally produce free radicals. Disease-causing organisms are destroyed by free radicals, which are used much like "bullets" by macrophages.[115,116] Free radical production is limited to the area around the leukocytes. If the local threshold of control is exceeded, free radicals migrate into adjacent tissues to produce inflammation.

Without antioxidant enzymes, we would die instantly. Depletion of antioxidants accelerates the aging process.[66,69,70] A clinical example of accelerated aging is the disease known as progeria, caused by hereditary absence of free radical protective enzymes. Within ten to fifteen years after birth, a victim of this disease proceeds through every aspect of the aging process, including wrinkled, dried, and sagging skin, baldness, bent and frail body, arthritis, and advanced cardiovascular disease. One form of progeria has been successfully treated by administering catalase, an enzyme that decomposes peroxides. Heredity determines each individual's unique resistance to free radical diseases, and each person correspondingly varies in tolerance to dietary and life-style abuse.

OXYGEN TOXICITY

Ongoing free radical pathology is often referred to as "oxidative stress." Ground state or unexcited atmospheric oxygen has the unique property of being both a free radical generator and a free radical scavenger.[66,70,117,118] Although a liter of normal atmospheric air on a sunny day contains over 1 billion hydroxyl free radicals,[90] oxygen at normal physiologic concentrations in living tissues neutralizes more free radicals than it produces.[119] When oxygen tension is reduced, as occurs in an ischemic organ, oxygen becomes a net contributor to free radical damage.[69,70,120]

Oxygen in high concentrations for prolonged periods of time can cause severe toxicity and death, primarily by free radical damage to the lungs and brain. Under proper conditions, intermittent oxygen administered for short periods in a hyperbaric (HBO) chamber can promote repair of free radical damage and stimulate an adaptive increase of superoxide dismutase.[118,121]

FREE RADICAL PROTECTION

Proper oxygenation enhances defenses against free radicals. Aerobic exercise stimulates blood flow and improves oxygen utilization, resulting in adequate oxygenation to remote capillary beds. When conditions of health exist, oxygen acts as a free radical scavenger during exercise and reduces free radical pathology.

The essential and desirable role played by nutritional trace elements as components of antioxidant enzyme systems has already been described. Each molecule of mitochondrial SOD contains three atoms of manganese. Each molecule of cytoplasmic SOD contains two atoms of zinc and one atom of copper. Each molecule of glutathione peroxidase contains four atoms of selenium. Catalase and peroxidase contain iron. Elemental selenium is also an antioxidant, independent of its function as an enzyme co-factor.[70]

The human body lacks an effective defense against the very destructive free radical precursor, excited state singlet oxygen. When superoxide radicals exceed the concentration that can be safely decomposed by superoxide dismutase (SOD), they spontaneously produce singlet oxygen. Polynuclear aromatic hydrocarbons (PAH) and aldehydes found in tobacco tar and tobacco combustion products can produce a variety of free radicals, often mediated by singlet oxygen. The most important protection against singlet oxygen is dietary intake of beta carotene (pro-vitamin A), the yellow pigment found in carrots and other fruits and vegetables.[122,123] Recent epidemiologic evidence correlates increased dietary beta carotene with reduced incidence of cancer, even in smokers. Fully active vitamin A lacks this protective activity.[69,70]

Vitamin E (tocopherol), vitamin C (ascorbate), selenium in glutathione peroxidase, the amino acid cysteine in reduced glutathione, riboflavin, and niacin are all interrelated in a recycling process that provides ongoing neutralization of free radicals. If each one of these nutrients is present in adequate amounts, they can all be restored to their active antioxidant forms after reacting with free radicals. The process proceeds as follows: Vitamin E neutralizes a free radical by being oxidized to tocopherol quinone. Tocopherol quinone is recycled to reduced vitamin E (tocopherol) by vitamin C, which is, in turn, oxidized to dehydroascorbate. Interestingly, the ratio of ascorbate to dehydroascorbate diminishes progressively with age and no specie of animal survives when that ratio falls below one to one.[70] Oxidized (dehydro) vitamin C is then recycled to ascorbate by glutathione peroxidase. Glutathione peroxidase is returned to its active form by oxidation of reduced glutathione. Oxidized glutathione is reduced by the riboflavin-dependent enzyme, glutathione reductase. Glutathione reductase is reactivated by niacin-dependent NADH. Oxidized NAD is metabolized in the electron transport system, and energy that originated as a free radical is utilized for desirable metabolic purposes. Subsequent steps in energy metabolism utilize a variety of trace element and vitamin dependent enzymes. From this stairstep cycle of oxidation-reduction reactions, it is obvious that continued activity of each component depends on an adequate supply of all components in the system.[92] This interdependence explains the often equivocal results obtained from clinical trails utilizing only one antioxidant alone.

If free radical production exceeds the capacity of this system, other antioxidant mechanisms are used. If those other mechanisms are not adequate, serious damage to cell membranes, protein molecules, and nuclear material results.[66,69,70,79,87,89,91]

An understanding of free radical defenses provides a rational for nutritional supplementation with vitamins and trace elements, in safe amounts and in proper physiologic ratios–including vitamin E, beta carotene, vitamin C, glutathione, and B-complex vitamins. Although high levels of most water-soluble vitamins are rapidly excreted, transient tissue elevations can enhance free radical protection and other metabolic functions.[70]

INCREASED PRODUCTION OF FREE RADICALS

When free radicals in living tissues exceed safe levels, the result is cell destruction, malignant mutation, tumor growth, damage to enzymes, and inflammation, which manifest clinically as age-related, chronic degenerative diseases. Each uncontrolled free radical has the potential to multiply by a million-fold.[66,69,70,79,87,89,91]

Dietary fats, especially those containing polyunsaturated fats, are the leading sources of pathological free radicals. Double-bonds on unsaturated fatty acids can combine with atmospheric oxygen, creating lipid peroxides.

Lipid peroxidation occurs when fats and oils are exposed to air. Oxidative damage to dietary fats and oils is catalyzed by metallic ions, especially iron and copper. For example, peanuts crushed to make peanut butter are rich in both iron and copper, which are released into the unsaturated oil when the peanut is disrupted. Iron and copper are potent catalysts of lipid peroxidation and increase the rate of rancidity of peanut oil by a million times. (Massively peroxidized lipids are called rancid; however, very extensive peroxidation can exist without a detectable rancid odor or taste.[70]) Extensive lipid peroxidation inevitably results during the manufacture of peanut butter.[70] The same is true of most other salad and cooking oils, even so-called cold-pressed oils.[70]

The richer the oil is in unsaturated fatty acids, the more readily peroxidation will occur. The rate of peroxidation is proportional to the square of the number of unsaturated bonds in each fatty acid molecule. Other factors that increase the rate of peroxidation of fats and oils include heat, atmospheric oxygen, light, and trace amounts of unbound metallic elements.[70,87] Oils prepared in the dark, at low temperatures, in an atmosphere of pure nitrogen, and containing fat-soluble antioxidants such as vitamin E, would be the safest for nutritional use.[70] Such oils are not commercially feasible.

Unsaturated vegetable oils containing iron and copper are routinely exposed to heat and oxygen when foods are fried. This creates the worst possible combination. Oils used in the manufacture of salad dressings, such as mayonnaise, contain a very high concentration of lipid peroxides. The poorest quality oils are customarily used because rancidity is masked by heavy seasoning.[70]

Peroxidation and hydrogenation of vegetable oils during the manufacture of margarine and shortening also results in cis to trans-isomerization. Trans-isomerization alters the three-dimensional configuration of dietary fatty acid constituents from their normal cis coils to straightened trans chains. Trans-fatty acids are incorporated into cell membranes in the place of naturally occurring cis forms, causing changes in membrane structure and impairing the function of phospholipid-dependent enzymes contained in cell membranes.[66,69,70,124,125] Substrate recognition by enzymes that synthesize cell membranes is not adequate to distinguish between these two stereo isomers.[70]

Phospholipids in cell membranes are damaged by free radicals. The process can be initiated by dietary peroxidized fats. Arachidonic and linoleic acids, both prostaglandin precursors, are lost in the process.[69,70] These losses can be measured by gas chromatography. Cell membranes containing trans-fatty acids have altered fluidity characteristics that increase membrane permeability and interfere with transport enzymes for sodium, potassium, calcium, magnesium, and other substances.[66,69,70] Adverse effects from trans-isomerization of fatty acids are additive to the other adverse effects of lipid peroxidation.[66,69,70]

Very little attention has been given to the quality of dietary fats and oils. The emphasis has mistakenly been placed on the ratio of saturated to unsaturated fatty acids, irrespective of lipid peroxidation and content of trans-isomers. We now know that the unsaturated fats are relatively more toxic.[70] Margarine contains far more peroxides and trans-fatty acids than butter, and limiting consumption of all fats and oils is desirable.[70]

The quantity of dietary fat is the most important factor in free radical pathology, not the ratio of unsaturated to saturated fatty acids.[66,69,70] If dietary fats and oils are consumed in fresh, whole unfractionated, unprocessed foods, they will not be peroxidized and will form healthy cell membranes with normal cis fatty acid configurations. They will also result in a normal balance of prostaglandins. Although fully saturated fats are not subject to lipid peroxidation, all animal fats contain some unsaturated fatty acids and cholesterol, both of which undergo peroxidation. Animal experiments have shown that if as little as 1 percent of dietary cholesterol is consumed in its oxidized form, atherosclerosis may result.[110,111,126]

How much fat can the human body tolerate? Evidence indicates that the average person can safely consume 20 percent of his or her dietary calories as commercially available fat without exceeding the limits of endogenous free radical protection.[69,70] The quality of dietary fats and oils can increase or decrease this percentage. The average American now consumes in excess of 40 percent of his or her calories as fat, mostly of poor quality, and with no consideration for rancidity. Lipid peroxidation can occur even when foods are frozen, although it then occurs more slowly.[70]

Lecithin is a phospholipid rich in unsaturated fatty acids. Analysis of all commercially available lecithin indicates a high degree of peroxidation. The only way to extract lecithin without extensive peroxidation is in an atmosphere of 100 percent nitrogen, which involves a prohibitively costly process.[70]

Research in senility, dementia, brain ischemia, stroke, and spinal cord injury provides a wealth of evidence to incriminate free radicals as a cause of nervous system disease and also provides a rationale for treatment. The brain and spinal cord contain the highest concentration of fat of any organ. Central nervous system fats are rich in highly unsaturated arachidonic and docosahexanoic acids. As previously stated, the rate of lipid peroxidation increases exponentially with the number of unsaturated double-bonds per fatty acid molecule. Docosahexanoic and arachidonic acids will peroxidize many times more readily than other lipids. It is therefore desirable for the central nervous system to have additional protection against free radicals. Vitamin C is highly concentrated in the brain by a metabolically active pump mechanism. Ascorbate is 100 times more concentrated in the central nervous system than in most other organs.[70] There are two ascorbate pump mechanisms in the series. The first increases the concentration ten-fold from blood to cerebrospinal fluid. A second pump mechanism extracts cerebrospinal fluid ascorbate and concentrates it by another factor of ten surrounding neurones of the brain and spinal cord. An important function of this vitamin C is protecting fats in nerve tissues from peroxidative damage. The disappearance rate of vitamin C from the central nervous system is used as an indicator of the rate of lipid peroxidation following ischemia or trauma.[127,128]

Observations of experimental spinal cord injury provide further support for treatments based on free radical protection. In animal experiments it has been shown that a minor contusion to the spinal cord will result in a very rapid breakdown of the unsaturated fatty sheaths surrounding nerve pathways. Even a small bruise causes capillaries to leak blood. Erythrocytes are quickly caught up in a fibrin clot, causing hemolysis and release of free iron and copper. These metals act as potent catalysts, which combine with oxygen to increase the rate of lipid peroxidation by up to a million-fold.[70] Animal experiments show that this chain reaction can be squelched in at least two ways: (1) The spinal cord can be irrigated with a potent free radical scavenger, such as dimethyl sulfoxide (DMSO), or (2) the iron and copper catalysts can be inactivated by bathing the injured area in a chelating solution containing EDTA or d-penicillamine.[70,77,108,129-131] Hyperbaric oxygen also appears to halt free radical progression, preserving spinal cord and brain function. Both DMSO and hyperbaric oxygen are now being utilized in research protocols at major medical centers to prevent permanent paralysis in victims of spinal cord and brain injuries. Results thus far indicate that if treatment is begun within the first thirty minutes, or at most within the first two hours, the outcome is much better than would otherwise have been expected.[132-147]

Chelation, hyperbaric oxygen, and dietary fat restriction have been observed to alleviate or temporarily reverse the progression of multiple sclerosis (MS).[34,148,149] MS victims experience degeneration of the unsaturated fatty insulating material of nerve pathways in the brain and spinal cord. Although free radical damage may be just one link in the chain of cause and effect, it has been possible to slow the progression of this devastating disease in many victims by using treatment principles that reduce free radical pathology.

TOBACCO AND ALCOHOL

It has long been known that habitual use of tobacco and excessive alcohol are related to illness and premature death. Alcohol causes damage and scarring of the liver and cancer in the mouth and digestive organs.[150] Alcohol is metabolized to acetaldehyde, which is a free radical generator. Acetaldehyde is closely related to formaldehyde (embalming fluid), which causes cross-linkages of connective tissue (tanning) by free radical reactions.[69,70,151,152] Cirrhosis of the liver might therefore be considered quite literally as a form of antemortem embalming.

Tobacco is associated with an increased incidence of atherosclerosis, cancer, and other diseases. Tobacco contains polynuclear aromatic hydrocarbons that produce free radicals and free radical precursors in its combustion tars. These substances put a great strain on the body's free radical defenses and accelerate the onset of cancer, atherosclerosis and other degenerative conditions.[153] Processed tobacco contains cadmium, a heavy metal ten times more toxic than lead, which also acts as a catalyst of free radical reactions and actively competes with zinc in metalloactivated enzymes.

CELL MEMBRANE METABOLISM

Every cell in the body is enveloped in a bipolar phospholipid membrane in which are suspended large enzyme molecules and other metabolically active components. This membrane has the characteristics of a viscous fluid and is constantly changing. It exhibits unidirectional permeability to substances that are not desired within the cell. The polar (water-soluble) ends of phospholipid molecules line the inner and outer surfaces of the cell membrane, while the nonpolar (fat soluble) tails point inward, traversing the interior of the cell wall. The normal cis configuration of phospholipids causes them to coil around proteins and other molecular constituents within the membrane. This cis curvature greatly influences the physical characteristics of the cell membrane and its associated enzyme activity.[66,69,70] Unoxidized cholesterol is widely disbursed within cell membranes and acts to protect against free radical damage.[70] Oxidized dietary cholesterol that has been altered in food processing offers no such antioxidant protection and is itself atherogenic.[70,110] When free radicals approach a cell cholesterol and a wide variety of other defenses normally prevent them from damaging the cell wall.

Large energy-dependent "pump" enzymes span the full thickness of cell membranes. They are bathed in plasma on the exterior and extend into the cytoplasm of the cell. One such enzyme maintains a higher concentration of sodium ions on the exterior of the cell wall and potassium on the interior. Another enzyme keeps a much higher concentration of calcium outside the cell and magnesium inside. Cellular organelles, including mitochondria, lysosomes, endoplasmic reticuli, and Golgi mechanisms, are also enveloped in bipolar lipid membranes containing numerous energy-dependent transport mechanisms. The mitochondrial membrane contains coenzyme Q, necessary for energy production. Mitochondria are the powerhouses of cells, and they normally produce free radicals during transport of electrons, in the process of oxidative phosphorylation. Adequate antioxidant protection prevents these free radicals from proliferating.[66,69,70]

Receptor sites on cell membranes for neurotransmitters, insulin, and various other hormones may be damaged by excessive free radicals. The calcium-magnesium and sodium-potassium pump mechanisms can be impaired, allowing excessive calcium and sodium to enter cells. Free radicals damage nuclear membranes, altering nuclear pores and genetic material, causing impairment of protein synthesis and cell replication. Free radical mutations of DNA may produce uncontrolled cell division leading to cancer. Free radicals increase the activity of guanylate cyclase, which stimulates cell multiplication. Lymphoid tissues are very rich in unsaturated fatty acids, and free radical damage may selectively cause immunologic abnormalities.[70] The immune system may then attack the body's own tissues in so-called autoimmune disease or it may be weakened and fail to recognize and destroy pathogenic organisms and malignant cells.[66,69,70,74]

CALCIUM METABOLISM

Free radical impairment of the calcium-magnesium pump mechanisms allows excessive calcium to enter the cell. Calcium activates phospholipase-A_2, which cleaves arachidonic acid from membrane phospholipids. Increased levels or arachidonic acid then produce prostaglandins and leukotrienes, creating more free radicals in the process.[69,70] Leukotrienes are potent inflammatory substances that attract leukocytes. Leukocytes, as noted previously, produce superoxide free radicals during phagocytosis. Leukocytes out of control, stimulated by leukotrienes, produce excessive free radicals and cause inflammatory damage to surrounding tissues.[69,70,154] Small capillaries and arterioles dilate, causing edema and leakage of erythrocytes through blood vessel walls. Platelets are stimulated to produce microthrombi. Erythrocytes hemolyze, releasing free copper and iron, which catalyze an increase of peroxidative damage to adjacent tissues.

Excessive calcium in smooth muscle cells, caused by free radical damage to cell walls, can be bound by calmodulin, which activates myosin kinase, which in turn phosphorylates myosin. Myosin and actin then constrict, causing the muscle cell to shorten. By this mechanism, increased calcium within arterial smooth muscle cells leads to spasm. The same occurs in cells of the myocardium. When muscle fibers encircling arteries constrict, blood flow is reduced. Calcium channel blockers relieve symptoms by slowing abnormal entry of calcium into cells, but they do not correct the underlying cause of the problem--free radical disruption of cell membranes.[69,70,155] Myocardial function is also impaired by the excessive intracellular calcium, reducing the efficiency of oxygen utilization and placing an extra burden on an already impaired coronary artery system. If a coronary (or other) artery is partially occluded by atherosclerotic plaque, a minor degree of spasm superimposed on the mechanical lesion may cause disabling symptoms. Myocardial infarction has been observed to occur as a result of pure spasm in a coronary artery that was completely free from plaque.[156] Thromboxane and serotonin are released by platelets in the presence of free radicals.[69,70] Thromboxane and serotonin are both potent mediators of arterial spasm.

Excessive concentrations of intracellular calcium results from a variety of other factors. Ionized plasma calcium, the metabolically active fraction not bound to protein, slowly increases with age, partially as a result of excessive dietary phosphates. The higher the concentration of ionized calcium outside a cell, the harder the calcium pump mechanism must work to prevent excessive calcium from leaking in. Nutritional calcium antagonists also slow calcium influx. These include dietary magnesium, manganese, and potassium. Magnesium and manganese intakes are suboptimal in the average American diet.[113] Diets are rarely deficient in potassium, but an excessively high ration of dietary sodium to potassium is common, allowing more sodium to diffuse into cells, poisoning metabolism in other ways. Losses of potassium and magnesium caused by diuretic therapy potentiate this problem.

The efficiency of energy metabolism is impaired by excessive stress. Stress increases circulating catecholamines that partially inhibit ATPase. Cells lose potassium and retain calcium and sodium under stress because of relative inhibition of both magnesium-calcium ATPase and sodium-potassium ATPase.

Catecholamines produce free radicals when they are metabolized.[69,70] If the nervous system's free radical defense capacity has been exceeded, stress-related catecholamines may cause free radical damage to neuronal receptors. This is one explanation for stress-related nervous disorders. Metabolic degradation products of dopamine, a catecholamine neurotransmitter, are capable of causing free radical damage to neuronal receptor sites in the brain, a possible cause of Parkinson's disease and some types of schizophrenia.[157] Neuronal receptors for norepinephrine can be damaged by free radicals, causing depression. Cardiac disease has also been shown to result from catecholamine induced free radicals.[158]

In recent animal experiments, primates subjected to stress were found to have an increased incidence of atherosclerosis, even while fed a diet that is otherwise protective against that disease. Increased free radical pathology associated with increased catecholamine metabolism could explain that finding.

Dementia of the Alzheimer's type is thought by some authorities to be the result of brain cell destruction by free radicals.[157] Arrest of the disease or improvement in the condition of significant percentage of patients with Alzheimer's dementia has been reported following treatment with deferoxamine, and iron chelating agent.[159] Iron is a potent catalyst of lipid peroxidation. Accumulations of aluminum, lipid, and protein breakdown pigments in areas of the brain affected by Parkinson's and Alzheimer's are now thought to be late events, not causative factors, much like the accumulation of calcium and cholesterol in the arterial plaque.

Before the significance of free radical pathology became known, it was hypothesized that EDTA chelation therapy had its major beneficial effect on calcium metabolism. It now seems apparent that calcium is just another link in the chain of cause and effect created by free radical damage. EDTA can influence calcium metabolism in many ways, but effects on calcium were never adequate by themselves to fully explain the many benefits observed following chelation. The lack of an acceptable scientific explanation has played a significant role in delaying general acceptance of EDTA chelation therapy.

EDTA lowers ionized plasma calcium during infusion. The body attempts to maintain homeostasis by producing an increase in circulating parathormone.[160] The intermittent three to four hour pulses of increased parathormone caused by EDTA infusion have a profound effect on bone metabolism.[161] Frost's recently proposed concept of bone metabolism, known as the Basic Multicellular Unit (BMU) theory, is now accepted by experts in this field.[162] The BMU theory helps to explain the causes and treatment of osteoporosis and osteopenia.

The BMU is a group of metabolically active cells that control the turnover of approximately 0.1 mm^3 of bone tissue. When a BMU is activated, it goes through a cycle consisting of an initial three to four weeks of bone absorption (osteoclastic phase) followed by a two-to-three-month period of bone reformation (osteoblastic phase). Net increase or decrease in bone density at the end of the entire three-to-four-month cycle depends on the rate and completeness of bone turnover. Hormone regulation of BMUs also includes growth hormone, thyroxin, and adrenal corticosteroids, but parathormone remains the most important controlling factor.[161]

Chronic excess concentration of parathormone has long been known to cause bone destruction, but brief pulsatile increases in parathormone levels, which occur during intravenous EDTA chelation therapy, result in an increase of new bone formation.[163]

Anabolic activation of BMUs by pulsatile parathormone secretions provides another possible explanation for the delayed improvement seen in chelation patients. Pathologic calcium deposits may be removed from arteries and other soft tissues for utilization in new bone formation. In their original studies, Meltzer and Kitchell reported treating ten severely handicapped men, all of whom suffered intractable angina, with EDTA. After approximately twenty infusions, therapy was discontinued because of initially disappointing results. Three months later nine out of ten patients returned to report marked relief of angina, despite no change in their life-styles, such as altering smoking or nutritional habits.[7] This three-month delay in achieving full benefits has remained a consistent observation by chelating physicians over the years.

Postmenopausal women who are not supplemented with estrogen experience a great increase in circulating follicular stimulating hormone (FSH). Elevated FSH interferes with new bone formation by BMU cells and is regarded as a possible cause of postmenopausal osteoporosis.[164]

It was previously hypothesized that the removal of calcium from plaque and from undesirable cross-linkage sites could explain most of the benefits seen from chelation. Cross-linkages of other types, such as disulfide bonds caused by free radical reactions, are also important. EDTA can correct abnormal disulfide bridging and cross-linkages caused by calcium, lead, cadmium, aluminum, and other metals. This reduction of cross-linkages can improve elasticity of vascular walls and other tissues.[36,37]

Vascular changes may not be seen arteriographically, in spite of significant clinical improvement. Using Poiseuille's Law of hemodynamics, one can demonstrate that with perfect laminar flow, a mere 19 percent increase in the diameter of a blood vessel will double the rate of blood flow. In a vessel with turbulent flow around an atherosclerotic plaque, this figure decreases to a less than 10 percent increase in diameter to cause a doubling of blood flow. In an organ with compromised circulation, an increase in blood flow of 10 to 20 percent could result in significant functional improvement, including alleviation of ischemic symptoms. Changes in diameter of such small magnitude cannot be detected on arteriograms.

EDTA CHELATION THERAPY

EDTA can reduce the production of free radicals by a million-fold.[70,109,165] It is not possible for free radical pathology to be catalystically accelerated by metallic ions in the presence of EDTA. Traces of unbound metallic ions are necessary for uncontrolled proliferation of free radicals in living tissues. EDTA binds ionic metal catalysts, making them chemically inert and removing them from the body. Concentrations of metallic ions with the ability to catalyze lipid peroxidation are so tiny that even the traces remaining in distilled water can initiate such reactions.[70,119]

Metallic ions with the ability to catalyze lipid peroxidation are those that readily change electrical valence by one unit. Two essential nutritional elements, iron and copper, are the most potent catalysts of lipid peroxidation. Catalytic iron and copper accumulate near phospholipid cell membranes, in joint fluid, and in cerebrospinal fluid with age and are released into tissue fluids following trauma or ischemia. These unbound extracellular iron and copper ions have been shown to potentiate free radical tissue damage.[79,87,89,91,166-171]

TOXIC HEAVY METALS

EDTA has long been accepted as a treatment of choice for heavy metal poisoning. Toxic heavy metals affected metabolism in a variety of ways. Poisonous metals such as lead, mercury, and cadmium react avidly with sulfur-containing amino acids on protein molecules. When lead reacts with sulfur on the cysteine or methionine moiety of an enzyme, its activity is inhibited or destroyed. Chelation therapy reactivates enzymes by removing toxic heavy metals. Concentrations of lead in human bones have increased more than five hundred-fold since the Industrial Revolution.[172] Bone lead is in equilibrium with other vital organs and tends to be released into the circulation under stress, increasing toxicity when it can least be tolerated.[173]

Lead takes on additional importance with respect to the antioxidant roles of glutathione and glutathione peroxidase. Lead reacts vigorously with sulfur-containing glutathione and prevents it from reacting with free radicals. As previously described, reduced glutathione is an essential antioxidant in the recycling of vitamins E and C, glutathione peroxidase, glutathione reductase, and NADH. Lead therefore cripples the free radical protective activity of that entire array of antioxidants.

Lead reacts with selenium even more avidly than with sulfur inactivating the selenium-containing enzyme, glutathione peroxidase. That enzyme protects against lipid peroxides in addition to performing its role in the antioxidant recycling system. Other toxic heavy metals also inactivate glutathione peroxidase. Multi element hair analysis is becoming accepted as a cost-effective screening test for heavy metal toxicity and also as a screening method for the nutritional status of several nutritional elements.[113,174-199] Definitive testing may then be selected, based on screening data from hair analysis.

The wide variety of benefits reported for EDTA chelation therapy can now be better understood. Copper and iron are not readily chelated by EDTA when tightly bound to metal containing enzymes and metal carrier molecules. On the peroxidation catalysts, they are quite readily removed by EDTA. Iron and copper both accumulate with age at pathological sites, where they catalyze free radical damage.[34,166-171] EDTA binds much more tightly to iron, copper, and other free radical catalysts than it does to calcium. EDTA will only bind calcium if none of the other ions is readily available.[34]

Iron accumulates more slowly in women during the childbearing years because of monthly menstrual losses. Younger women also have significant protection against atherosclerosis. That protection is lost at menopause. Body iron stores, best reflected by serum ferritin, accumulate in men four times more rapidly than in premenopausal women.[200] The risk of atherosclerosis is four times greater in men in this same age group. Data from the Framingham study document immediate loss of protection and, following hysterectomy, reversion of women to approximately the same relative risk of atherosclerosis as men, even if the ovaries are not removed.[201-203] These observations indicate that slower iron accumulation rather than endocrine influence is primarily responsible for reduced atherosclerosis in premenopausal women.[204] The fact that iron is a potent catalyst of lipid peroxidation provides an explanation for these clinical and epidemiologic findings. EDTA has a very high affinity for unbound iron.

The affinity of EDTA for various metals, at physiologic pH, in order of decreasing stability is seen below. In the presence of a more tightly bound metal, EDTA releases other metals lower in the series and will chelate the metal for which it has a greater affinity.[205] Calcium is near the bottom of the list, while iron, copper, and toxic metals are near the top.

chromium^{2+}
iron^{3+}
mercury^{2+}
copper^{2+}
lead^{2+}
zinc^{2+}

cadmium^{2+}
cobalt^{2+}
aluminum^{3+}
iron^{2+}
manganese^{2+}
calcium^{2+}
magnesium^{2+}

Magnesium is a calcium antagonist, relatively deficient in many chelation patients, and is the metallic ion least likely to be removed by EDTA. In fact, EDTA is best administered as magnesium-EDTA, providing an efficient delivery system that increases magnesium stores.

Free radical inhibition by EDTA may explain the recently published observation of Blumer in Switzerland, who reported a 90 percent reduction in deaths from cancer in a large group of chelation patients who were chelated and who had been carefully followed over an eighteen-year period. When compared with a statistically matched control group, Blumer reported a ten times greater death rate from cancer in the untreated group, compared to the death rate of patients who had been treated with EDTA.[38] A greatly reduced incidence of cardiovascular deaths was also observed. The common denominator of both cancer and atherosclerosis is now believed to be free radical pathology.[66,69,70] Blumer used calcium-EDTA, which is assumed to have precluded any direct effect on calcium metabolism in these patients. Although he attributed benefits to lead removal, the patients studied had been exposed to the same environmental lead as an average resident in any large city. Chelation and removal of free radical catalysts seem a more likely explanation. It was Demopoulos[89] who first proposed that chelation be used to control free radical pathology. Demopoulos also pointed out that many antioxidants have chelating properties.[66,89]

EDTA increases the efficiency of mitochondrial oxidative phosphorylation and improves myocardial function, quite independently or any effect on arterial blood supply.[206] Treatment with deferoxamine, an iron chelator, has been shown to improve cardiac function in patients with increased iron stores.[207] Chelation with deferoxamine has also been shown to reduce inflammatory responses in animal experimentation.[208] Sullivan has suggested that periodic donation of blood be studied as a means to reduce the risk of atherosclerosis in men and postmenopausal women.[204]

By interrupting the process of free radical damage, EDTA chelation therapy can promote normal healing. Slow healing of damaged tissues provides another explanation for the time lapse of several months following chelation before full benefit is observed. Stimulation of normal healing seems far superior to the mere suppression of symptoms achieved with most other therapies.

CHELATION AND ATHEROSCLEROSIS

Speaking briefly of the cause of occlusive atherosclerotic cardiovascular disease, we must mention that it is known that if an injury results in bleeding, homeostatic mechanisms must quickly stop the flow of blood to prevent hemorrhage and death. This regulation of blood loss is under the control of a variety of mechanisms, including hormones, the prostaglandins. Prostaglandins are produced and degraded continuously and very rapidly in endothelial cells and platelets. Prostaglandins have a half-life measured in seconds and must be constantly synthesized at a controlled rate and with a proper ratio between various types to maintain normal blood flow.

The two most important prostaglandins in relation to blood flow and atherosclerosis are prostacyclin and thromboxane. Prostacyclin reduces the adhesiveness of platelets, allowing free flow of blood cells and plasma, reducing the tendency to fibrin deposition and thrombi. Prostacyclin relaxes encircling muscle fibers in artery walls reducing spasm. Thromboxane does the opposite. It causes intense spasm in blood vessel walls and stimulates platelets to adhere.[209] In oversimplified terms, thromboxane may be seen as undesirable and prostacyclin as desirable. In actual fact, a proper balance must be maintained to protect against injury and hemorrhage, on the one hand, and to maintain normal circulation, on the other.

Synthesis of prostacyclin is greatly inhibited by lipid peroxides and free radicals, while thromboxane production remains unaffected. If lipid peroxides are present, either from dietary intake of peroxidized fats and oils or from nearby peroxidation of lipoprotein cell membranes, less prostacyclin is produced to balance the effects of thromboxane.[69,70,210]

Damage to vascular endothelium occurs repeatedly from free radicals and from minor hemodynamic stresses related to blood flow during daily physical activity. Some such damage may be caused by disordered immunity. Under healthy circumstances, minor vascular injuries are rapidly healed, aided by a thin layer of platelets, which coat the disrupted surface with a protective blanket.[111] If free radical protection is inadequate because local control capacities have been exceeded, the resulting proliferation of free radicals blocks the production of prostacyclin. Without prostacyclin, thromboxane is unopposed and causes the injured area of the arterial wall to become excessively attractive to platelets and the platelets to become attracted to each other. Platelets deposit in an abnormally thick aggregation. This growing layer of platelets traps a number of leukocytes, which also produce free radicals. A network of fibrin and microthrombi is formed, and erythrocytes become trapped. Some of these erythrocytes hemolyze, causing iron and copper to be released into the surrounding area. These metallic catalysts produce an explosive increase in free radical oxidation of cholesterol and phospholipid in nearby cell membranes. Prostacyclin production is inhibited for some distance along the blood vessel.

The resulting high concentration of free radicals can damage nuclear material in cells of the artery causing mutation and loss of control of cell replication. Lipid peroxides increase the activity of guanylate cyclase, which speeds mitosis.[66,69,70] This sequence of events eventually produces an atheroma, an enlarging tumor consisting of mutated, rapidly multiplying, multipotential cells that have lost their former high degree of differentiation and specialized function. Atheroma cells produce substantial amounts of connective tissue, collagen, and elastin. They act as macrophages, ingesting cellular debris. The monoclonal theory of atheroma formation first proposed by Benditt[211] most accurately fits the known facts. Cholesterol is oxidized by free radical activity, and some of the cholesterol oxidation products ingested by atheroma cells have vitamin D activity.[110]

We have described how intracellular calcium concentrations are abnormally high because of free radical damage to homeostatic mechanisms in cell walls. Localized excesses of vitamin D activity caused by oxidation of cholesterol can produce further calcium accumulations. Calcium and cholesterol deposits accumulate in the late stages of atheroma formation. Increasing amounts of cholesterol may be produced in an attempt to protect against further free radical damage. Some is even produced within the atheroma.[212] More cholesterol is oxidized by free radicals and oxidized cholesterol and cholesterol esters deposit within the plaque. Eventually the plaque expands to exceed its own blood supply. When the interior of the plaque grows too far from the closest intact circulation to receive adequate oxygen and other nutrients, the central core of the plaque degenerates into an amorphous fibro-fatty mass that contains varying amounts of calcium, cholesterol, connective tissue, and cellular debris. This necrotic core can ulcerate and become embolic. Free radical reactions continue to suppress prostacyclin synthesis, causing a continuous aggregation of platelets. These platelets release high concentrations of thromboxane and serotonin, promoting arterial spasm, which further occludes blood flow.

Symptoms of ischemia begin to occur when a blood vessel becomes approximately 75 percent occluded. A large meal, rich in peroxidized fats, can cause a sudden free radical insult, triggering an abrupt increase in spasm or even an acute thrombosis, resulting in an infarction of an ischemic organ.

Similar degenerative changes may occur in any part of the body. Cells swell and die as membranes become leaky and damaged. Cell membrane pump mechanisms become uncoupled or disabled. DNA damage results in mutations, atheroma, and cancer, increasing the probability of uncontrolled cell replication.[66,69,70] Lymphoid tissues and cells of the immune system become damaged.[66,69,70,74] Cross-linkages occur within connective tissue, elastin, and on enzyme molecules, caused by free radical reactions and metallic ions. Tissues age rapidly and organ functions deteriorate. Joints become hypertrophic, inflamed, and deformed. Leukotriene production and prostaglandin imbalances can cause arthritis and inflammatory change in other organs. Lysosomes rupture, releasing proteolytic enzymes, which can devastate cell contents. Lysosomes have been called the cells' digestive organs and, when disrupted, can lead to cellular autodigestion. Free radicals react with available selenium, increasing the excretion of selenium products, and can create a relative selenium deficiency. Cancer patients excrete selenium in amounts up to five times the normal rate, at a time when they need it the most.[113]

Antibody production and cellular immunity are impaired by free radicals. Cells of the immune system are especially rich in unsaturated fats and are therefore vulnerable to free radical pathology. Oxidized cholesterol and lipid peroxides are potent immunosuppressants.[69,70,74,110] Antigenic substances and malignant change, which would normally be protected against, can overwhelm a weakened immune system that has lost its reserve. Partially digested food proteins, which enters the circulation through the digestive tract, are no longer tolerated.[213-215] Adverse reactions to specific foods (so-called "food allergies") can develop. Free radical reactions, which occur normally in macrophages in the process of phagocytosis of antigens, can proceed out of control and cause tissue inflammation. Adverse tissue, reactions to a variety of nutritious foods and other environmental exposures are an increasingly common cause of illness. Avoidance of sensitizing foods and other triggering factors then becomes necessary to control symptoms.[216-220] Antigenic properties of and toxics produced by the common yeast, Candida albicans, normally present in the body in small numbers, may overwhelm the immune system. Candida organisms multiply more easily and produce potent toxins.[221-223] Yeast-related illness is potentiated in a population raised on antibiotics, birth control pills, adrenocorticosteroids, and a diet high in refined carbohydrates– all of which stimulate excessive growth of Candida albicans. Recent clinical reports indicate that the common yeast, Candida albicans, is an important cause of symptoms affecting a large percentage of the population.[224-227] A struggling immune system may become over reactive in other areas, attacking healthy tissues leading to autoimmune states.

TREATMENT AND PREVENTION OF FREE RADICAL PATHOLOGY

The development of cancer often takes decades from the initiating even to the onset of symptoms. If cancer-promoting factors are removed, free radical damage can be repaired and health can be restored by applied clinical nutrition, antioxidant therapy, and life-style corrections. Malignant cells in their early stages are able to undergo reverse transformation to a normal state. For example, smokers who stop the use of tobacco have approximately the same risk of cancer ten years later as those who never smoked.[70] Atherosclerosis involves a nonmalignant tumor, an atheroma, somewhat analogous to cancer. It could similarly regress with time, if causative factors are removed. Free radical pathology is the common denominator for atherosclerosis and cancer.

(1) Diet

Dietary fats and oils should be limited to 20 percent or less of total calories.[69,70] Consumption of fats and lipids that have been processed, extracted, exposed to air, heated, hydrogenated, or in any way altered and removed from the food in which they naturally occur should be reduced as much as practical. Consumption of carbohydrates that are depleted of trace elements (white flour, white rice, sugar) should be minimized. Total caloric intake should be moderated to maintain weight within 20 percent of ideal body weight. The use of table salt should be restricted. Diets should contain ample amounts of fiber-rich whole grains and fresh vegetables. Patients suffering with extensive free radical diseases should be stricter with diet until improvement occurs. Clinical improvement involves a healing process requiring months or years to complete.

(2) Nutritional Supplements

A scientifically balanced regimen of supplemental nutrients reinforces endogenous antioxidant defenses. Supplemental antioxidants and vitamins should include vitamins E, C, B_1, B_2, B_3, B_6, B_{12}, pantothenate, PABA, beta carotene, and glutathione. A balanced program of mineral and trace element supplementation should include magnesium, zinc, selenium, manganese, and chromium.[113] Trace elements can be toxic if taken to excess and copper and iron supplementation may speed free radical damage. Iron and copper should be supplemented only to treat deficiency states diagnosed by serum ferritin and whole blood or erythrocyte copper levels.[204,228] Trace element supplementation should be under the supervision of a health care professional who is knowledgeable in nutrition. Dietary histories and biochemical testing may allow trace element supplementation to be tailored to the needs of each individual. Hair analysis is a cost effective screening test for copper deficiency and excess as well as for the evaluation of several other nutritional elements and accumulations of toxic heavy metals. Hair is not a reliable indicator of iron stores.[113,174-199,228,229]

(3) Moderating Health-Destroying Habits

Tobacco: It is best to eliminate the use of tobacco altogether, but, if that is not possible, a marked reduction in exposure would be helpful. This applies to cigarettes, pipe tobacco, cigars, snuff, and chewing tobacco. Tobacco causes problems, even without combustion. Free radical precursors are absorbed from tobacco through the lining of the mouth and nose, even without inhaling smoke. A relatively healthy adult with adequate dietary intake of antioxidants may tolerate up to ten low-tar cigarettes (0.1 to 1 mg tar) per day without an increased risk of cancer, but even this amount increases the risk of atherosclerosis.[230]

Alcohol: Many victims of chronic degenerative diseases discover for themselves that alcohol is not well tolerated. For individuals with chronic illness, complete avoidance may be advisable. A healthy adult should be able to tolerate and detoxify one to two ounces of pure ethanol per twenty-four hours (four eight-ounce glasses of beer, four small glasses of wine, or two to three shot glasses of hard liquor at most).That dosage of alcohol can normally be consumed in twenty-four hours without the drinker's exceeding his or her capacity to neutralize the resulting free radicals.[69,70]

(4) Physical Exercise

Moderate physical exercise, even a brisk forty-five-minute walk several times per week, will help to maintain efficient utilization of oxygen. More vigorous aerobic exercise results in proportionately greater benefits. Lactate accumulates up to twice the normal levels in tissues during endurance exercise, even during a brisk walk.[231] Lactate has proven chelating properties, and some benefits of exercise are thought to result from chelation of undesired metallic elements.[37]

(5) Hyperbaric Oxygen

Intermittent exposures to 100 percent oxygen at up to twice the normal atomospheric pressure can interrupt free radical pathology, restoring free radical control capacity in a variety of ways. Oxygen at reduced levels, as occurs in ischemic organs, causes free radical reactions to proceed more rapidly. Hyperbaric oxygen (HBO) raises oxygen tension in ischemic tissues to normal levels. HBO stimulates an adaptive increase in cellular levels of free radical protective enzymes, such as superoxide dismutase.[118] Rebound dilatation of blood vessels follows HBO, improving blood flow to ischemic organs. HBO helps to kill disease-causing organisms, especially anaerobic bacteria, and stimulates the ingrowth of new blood vessels to ischemic areas. HBO protects the fatty insulating sheaths surrounding nerve tracts in the brain and spinal column from free radical damage, relieving symptoms of stroke, senility, multiple sclerosis, and spinal cord injury. HBO is most effective in the early stages of these conditions.[121,148,232]

(6) EDTA Chelation Therapy

Removing toxic heavy metals and abnormally situated iron and copper with EDTA, even when present in ultratrace amounts, can arrest the progress of free radical damage. Other benefits occur from uncoupling of disulfide and metallic cross-linkages, by normalization of calcium metabolism, by reactivation of enzymes poisoned by lead and other toxic metals, and by restoration of normal prostacyclin production along blood vessel walls. Lasting benefits are possible following a series of intravenous EDTA infusions in conjunction with other preventive, nutritional, and therapeutic measures.

The proper scientific application of EDTA chelation therapy, as taught by the American Board of Chelation Therapy,[233-235] should lead to widespread recognition and acceptance of this well-documented, safe, and effective therapy.

REFERENCES

1. Clarke NE, Clarke CN, Mosher RE: The "in vivo" dissolution of metastatic calcium: An approach to atherosclerosis. *Am J Med Sci* 1955;229:142-49.
2. Schroeder HA, Perry HM, Jr.: Antihypertensive effects of metal binding agents. *J Lab Clin Med* 1955;46:416.
3. Clarke NE, Clarke CN, Mosher RE: Treatment of angina pectoris with disodium ethelyne diamine tetraacetic acid. *Am J Med Sci* 1956;232:654-66.
4. Boyle AJ, Casper JJ, McCormick H, *et al*: Studies in human and induced atherosclerosis employing EDTA. (Swiss, Basel) *Bull Schweiz Akad Med Wiss* 1957;13:408.
5. Muller SA, Brunsting LA, Winkelmann RK: Treatment of scleroderma with a new chelating agent, edathamil. *Arch Dermatol* 1959;80:101.
6. Clarke NE, Sr: Atherosclerosis, occlusive vascular disease and EDTA. *Am J Cardiol* 1960;6:233-36.
7. Meltzer LE, Ural ME, Kitchell JR: The treatment of coronary artery disease with disodium EDTA, in Seven MJ, Johnson LA (eds.): *Metal Binding in Medicine*. Philadelphia, J.B. Lippincott Co., 1960, pp 132-36.
8. Peters HA: Chelation therapy in acute, chronic and mixed porphyria, in Seven MJ, Johnson LA (eds.): *Metal Binding in Medicine*. Philadelphia, J.B. Lippincott Co., 1960, pp 190-99.
9. Seven MJ, Johnson LA (eds.): *Metal Binding in Medicine: Proceedings of a Symposium Sponsored by Hahnemann Medical College and Hospital,* Philadelphia. Philadelphia, J.B. Lippincott Co., 1960.
10. Kitchell JR, Meltzer LE, Seven MJ: Potential uses of chelation methods in the treatment of cardiovascular diseases. *Prog Cardiovasc Dis* 1961;3:338-49.

11. Peters HA: Trace minerals, chelating agents and the porphyrias. *Fed Proc* 1961;20(3) (Part II) (suppl 10):227-34.
12. Boyle AJ, Clarke NE, Mosher RE, McCann DS: Chelation therapy in circulatory and sclerosing diseases. *Fed Proc* 1961;20(3) (Part II) (suppl 10):243-57.
13. Soffer A, Toribara T, Sayman A: Myocardial responses to chelation. *Br Heart J* 1961 Nov 23:690.
14. Peripheral Flow Opened Up. *Medical World News,* Mar 15, 1963;4:36-39.
15. Boyle AJ, Mosher RE, McCann DS: Some in vivo effects of chelation—I: Rheumatoid arthritis. *J Chronic Dis* 1963;16:325-28.
16. Aranov DM: First experience with the treatment of atherosclerosis patients with calcinosis of the arteries with trilon-B (disodium salt of EDTA). *Klin Med* (Russ, Moscow) 1963;41:19-23.
17. Soffer A, Chenoweth M, Eichhorn G, Rosoff B, Rubin M, Spencer H; *Chelation Therapy,* Springfield, Illinois, Charles C. Thomas, 1964.
18. Soffer A: Chelation therapy for cardiovascular disease, in Soffer A (ed.): *Chelation Therapy*, Springfield, Illinois, Charles C. Thomas, 1964, pp 15-63.
19. Lamar CP: Chelation therapy of occlusive arteriosclerosis in diabetic patients. *Angiology* 1964;15:379-94.
20. Friedel W. Schulz FH, Schoder L: Therapy of atherosclerosis through mucopolysaccarides and EDTA (ethylene diamine tetraacetic acid). (German) *Deutsch Gesundh* 1965;20:1566-70.
21. Lamar CP: Chelation endarterectomy for occlusive atherosclerosis. *J Am Geriatr Soc* 1966;14:272-93.
22. Birk RE, Rupe CE: The treatment of systemic sclerosis with EDTA, pyridoxine and reserpine. *Henry Ford Hospital Medical Bulletin* 1966 June;14:109-39.
23. Lamar CP: Calcium chelation of atherosclerosis, nine years' clinical experience. Read before the Fourteenth Annual Meeting of the American College of Angiology, San Juan, PR, Dec 8, 1968. (Transcript available AAMP, 6151 West Century Blvd., #1114, Los Angeles, CA 90045).
24. Olwin JH, Koppel JL: Reduction of elevated plasma lipid levels in atherosclerosis following EDTA chelation therapy. *Proc Soc Exp Biol Med* 1968;128:1137-39.
25. Leipzig LJ, Boyle AJ, McCann DS: Case histories of rheumatoid arthritis treated with sodium or magnesium EDTA. *J Chronic Dis* 1970;22:553-63.
26. Brucknerova O, Tuläcek J: Chelates in the treatment of occlusive atherosclerosis. (Czechoslovakian, Praha) *Vnitr Lek* 1972;12:137-39.
27. Nikitina EK, Abramova MA: Treatment of atherosclerosis patients with Trilon-B (EDTA). (Russian, Moscow) *Kardiologiia* 1972;12:137-39.
28. Evers R: Chelation of vascular atheromatous disease. *Journal International Academy Metabology* 1972;2:51-53.
29. Kurliandchikov VN: Treatment of patients with coronary arteriosclerosis with unithiol in combination with vitamins. (Russian, Kiev) *Vrach Delo* 1973;6:8.
30. Zapadnick VI, *et al*: Pharmacological activity of unithiol and its use in clinical practice. (Russian, Kiev) *Vrach Delo* 1973;8:122.
31. David O Hoffman SP, Sverd J, Clark J, Voeller K: Lead and hyperactivity, behavioral response to chelation: A pilot study. *Am J Psychiatry* 1976;133:1155-58.
32. Gordon GB, Vance RB: EDTA chelation therapy for atherosclerosis: History and

Mechanisms of action. *Osteopathic Annals* 1976;4:38-62.
33. Proceedings: Hearing on EDTA Chelation Therapy of the Ad Hoc Scientific Advisory Panel on Internal Medicine of the Scientific Board of the California Medical Society, March 26, 1976, San Francisco, California. (Transcript available from AAMP, 6151 West Century Blvd., #1114, Los Angeles, CA 90045).
34. Halstead BW: *The Scientific Basis of EDTA Chelation Therapy*. Colton, CA, 1979. Golden Quill Publishers, Box 1278, Colton, CA 92324, 1979.
35. Grumbles LA: Radionuclide studies of cerebral and cardiac circulation before and after chelation therapy. Read before the American Academy of Medical Preventics, Chicago, Illinois, May 27, 1979. (Transcript available from AAMP, 6151 West Century Blvd., #1114, Los Angeles, CA 90045).
36. Bjorksten J: The cross-linkage theory of aging as a predictive indicator, *Rejuvenation* 1980;8:59-66.
37. Bjorksten J: Possibilities and limitations of chelation as a means for life extension. *Rejuvenation* 1980;8:67-72.
38. Blumer W, Reich T: Leaded gasoline–a cause of cancer. *Environmental International* 1980;3:465-71.
39. Carpenter DG; Correction of biological aging. *Rejuvenation* 1980;8:31-49.
40. Casdorph HR: EDTA chelation therapy, efficacy in arteriosclerotic heart disease. *J Holistic Med* 1981;3(1):53-59.
41. Casdorph HR: EDTA chelation therapy II, efficacy in brain disorders. *J Holistic Med* 1981;3(2):101-17.
42. McDonagh EW, Rudolph CJ, Cheraskin E: An oculocerebrovasculometric analysis of the improvement in arterial stenosis following EDTA chelation therapy. *J Holistic Med* 1982;4(1):21-23.
43. Olwin JH: EDTA Chelation Therapy. Read before the American Holistic Medical Association, University of Wisconsin, La Crosse, Wisconsin, May 28, 1981. (Available on audio cassette, AHMI, 6932 Little River Turnpike, Annandale, Virginia 22003).
44. McDonagh EW, Rudolph CJ, Cheraskin E: The influence of EDTA salts plus multivitamin-trace mineral therapy upon total serum cholesterol/high-density lipoprotein cholesterol. *Medical Hypothesis* 1982;9:643-46.
45. McDonagh EW, Rudolph CJ, Cheraskin E: The effect of intravenous disodium ethylenediaminetetraacetic acid (EDTA) upon blood cholesterol in a private practice environment. *Journal of the International Academy of Preventive Medicine* 1982;7:5-12.
46. Williams DR, Halstead BW: Chelating agents in medicine. *J. Toxicol: Clin Toxicol* 1983;19(10:1081-115.
47. Casdorph HR, Farr CH: EDTA chelation therapy III: Treatment of peripheral arterial occlusion, an alternative to amputation. *J Holistic Med* 1983;5(1):3-15.
48. Doolan PD, Schwartz SL, Hayes JR, Mullen JC, Cummings NB: An evaluation of the nephrotoxicity of ethylendiaminetetraacetate and diethlenetriaminepentaacetate in the rat. *Toxicol Appl Pharmacol* 1967;10:481-500.
49. Ahrens FA, Aronson Al: A comparative study of the toxic effects of calcium and chromium chelates of ethylendiaminetetraacetate in the dog. *Toxicol Appl Pharmacol* 1971;18:10-25.

50. Feldman EB: EDTA and angina pectoris. *Drug Therapy* 1975 Mar:62.
51. Wedeen RP, Mallik DK, Batuman V: Detection and treatment of occupational lead nephropathy. *Arch Intern Med* 1979;139:53-57.
52. Moel DI, Kuman K: Reversible nephrotoxic reactions to a combined 2, 3-dimercapto-1-propanol and calcium disodium ethylenediaminetetraacetic acid regimen in asymptomatic children with elevated blood levels. *Pediatrics* 1982;70(2):259-62.
53. McDonagh EW, Rudolph CJ, Cheraskin E: The effect of EDTA chelation therapy plus supportive multivitamin-trace mineral supplementation upon renal function: A study in serum creatinine. *J Holistic Med* 1982;4:146-151.
54. Cranton EM: Kidney effects of ethylene diamine tetraacetic acid (EDTA): A literature review. *J Holistic Med* 1982;4:152-157.
55. Batuman V, Landy E, Maesaka JK, Wedeen RP: Contribution of lead to hypertension with renal impairment. *N Engl J Med* 1983;5(2):871-879.
56. McDonagh EW, Rudolph CJ, Cheraskin E: The effect of EDTA chelation therapy plus supportive multivitamintrace mineral supplementation upon renal function: A study in blook urea nitrogen (BUN). *J Holistic Med* 1983;5(2):871-879.
57. Kitchell JR, Palmon F, Aytan N, Meltzer LE: The treatment of coronary artery disease with disodium EDTA, a reappraisal. *Am J Cardiol* 1963;11:501-6.
58. Cranton EM, Frackelton JP: Current status of EDTA chelation therapy in occlusive arterial disease. *J Holistic Med* 1982;4:24-33.
59. Wartman A, Lampe TL, McCann DS, Boyle AJ: Plaque reversal with MgEDTA in experimental atherosclerosis: Elastin and collagen metabolism. *J Atheros Res* 1967;7:331.
60. Wissler RW: Principles of the pathogenesis of atherosclerosis, in Braunwald E (ed): Heart Disease. Philadelphia, W.B. Saunders Col. 1980, pp 1221-36.
61. Kjeldsen K, Astrup P, Wanstrup J: Reversal of rabbit atherosclerosis by hyperoxia. *J Atheros Res* 1969;10:173.
62. Vesselinovitch D, Wissler RW, Fischer-Dzoga K, Hughes R, DuBien L: Regression of atherosclerosis in rabbits. I. Treatment with low fat diet, hyperoxia and hypolipidemic agents. *Atherosclerosis* 1974;19:259.
63. Sincock A: Life extension in the rotifer by application of chelating agents. *J Gerontol* 1975;30:289-93.
64. Wissler, RW, Vesselinovitch D: Regression of atherosclerosis in experimental animals and man. *Mod Concepts Cardiovasc Dis* 1977;46:28.
65. Walker F: The effects of EDTA chelation therapy on plaque, calcium, and mineral metabolism in arteriosclerotic rabbits, Ph.D thesis. Texas State University, 1980. (Available from University Microfilm International, Ann Arbor MI 48016).
66. Demopoulos HB, Pietronigro DD, Flamm ES, Seligman ML: The possible role of free radical reactions in carcinogenesis. *Journal of Environmental Pathology and Toxicology* 1980;3:273-303.
67. Harman D: The aging process. *Proc Natl Acad Sci USA* 1981;78:7124-28.
68. Dormandy TL: An approach to free radicals. *Lancet* 1983;ii:1010-14
69. Demopoulos HB, Pietronigro DD, Seligman ML: The development of secondary pathology with free radical reactions as a threshold mechanisms. *Journal of the American College of Toxicology* 1983;2(3):173-84.

70. Demopoulos HB: Molecular oxygen in health and disease. Read before the American Academy of Medical Preventics Tenth Annual Spring Conference, Los Angeles, California May 21, 1983. (Available on three audio cassettes from Instatape, P.O. Box 1729, Monrovia, CA 91016).
71. Ames BN: Dietary carcinogens and anticarcinogens. *Science* 1983;221:1256-64.
72. Dormandy TL: Free radical reaction in biological systems. *Ann R Coll Surg Engl* 1980;62:188-94.
73. Dormandy TL: Free radical oxidation and antioxidants. *Lancet* 1978;8:647-50.
74. Levine SA: Reinhardt JH: Biochemical-pathology initiated by free radicals, oxidant chemicals, and therapeutic drugs in the etiology of chemical hypersensitivity disease. *Journal of Orthmolecular Psychiatry* 1983;12(3):166-83.
75. Del Maestro RF: An approach to free radicals in medicine and biology. *Acta Physiol Scand* 1980;492(suppl.):153-68.
76. Poole CP: *Electron Spin Resonance. A Comprehensive Treatise on Experimental Techniques.* New York, Interscience Publishers, 1967.
77. Demopoulos HB, Flamm ES, Seligman ML, Mitamura JA, Ransohoff J: Membranes perturbations in central nervous system injury: Theoretical basis for free radical damage and a review of the experimental data, in Popp AJ, Bourke LR, Nelson LR, Kimelbert HK (eds).): *Neural Trauma.* New York, Raven Press, 1979;63-78.
78. Seligman ML, Mitamura JA, Shera N, Demopoulos HB: Corticosteroid (methylprednisolone) modulation of photoperoxidation by ultraviolet light in lipsomes. *Photochem Photobiol* 1979;29:549-58.
79. Pryor WA: Free radical reactions and their importance in biochemical systems. *Fed Proc* 1973;32:1862-69.
80. Pryor WA (ed): *Free Radicals in Biology. Volumes 1-3.*New York, Academic Press, 1976.
81. Lambert L, Willis ED: The effect of dietary lipid peroxides, sterols and oxidized sterols on cytochrome P-450 and oxidative demethylation. *Biochem Pharmacol* 1977a;26:1417-21.
82. Lambert C, Willis ED: The effect of dietary lipids on 3, 4, benzo(a)pyrene metabolism in the hepatic endoplasmic reticulum. *Biochem Pharmcol* 1977b;26: 1423-77.
83. Fedorenko VI: Effect of cysteine, glutathione and 1-p-chlorophenyltetrazolethione-2 on postradiation changes in the metabolic free radical content of albino rat tissues. *Radiobiologiia* 1979;19:67-73.
84. Fridovich I: Superoxide dismutases. *Annu Rev Biochem* 1975;147-59.
85. Black HS, Chan JT: Experimental ultraviolet light-carcinogenesis. *Photochem Photobiol* 1977;26:183-89.
86. Eaton GJ, Custer P, Crane R: Effects of ultraviolet light on nude mice: Cutaneous carcinogenesis and possible leukomogenesis. *Cancer* 1978;42:182-88.
87. Tappel AL: Lipid peroxidation damage to cell components. *Fed Proc* 1973;32: 1870-74.
88. Kotin P, Falk Hal: Organic peroxides, hydrogen peroxide, epoxides and neoplasia. *Radiat Res* 1963;3(suppl):193-211.

89. Demopoulos HB: Control of free radicals in the biologic systems. *Fed Proc* 1973;32:1903-08.
90. Walling C: Forty years of free radicals, in Pryor WA (ed): *Organic Free Radicals.* Washing. D.C., American Chemical Society, 1978.
91. Demopoulos HB: The basis of free radical pathology. *Fed Proc* 1973;32:1859-61.
92. Tappel AL: Will antioxidant nutrients slow aging processes? *Geriatrics* 1968;23: 97-105.
93. Coon MJ: Oxygen activation in the metabolism of lipids, drugs and carcinogens. *Nutr Rev* 1978;36:319-28.
94. Coon MJ: Reconstitution of the cytochrome P-450-containing mixed-function oxidase system of liver microsomes. *Methods Enzymol* 1978;52:200-06.
95. Coon MK, van der Hoeven TA, Dahl SB, Haugen DA: Two forms of liver microsomal cytochrome P-450, P-4501m2 and P-450M4 (rabbit liver). *Methods Enxymol* 1978, 52:109-17.
96. Panganamala RV, Sharma HM, Sprecher H, Geer JC, Cornwell DG: A suggested role for hydrogen peroxide in the biosynthesis of prostaglandins. *Prostaglandins* 1974;8:3-11.
97. Maisin JR, Decleve A, Gerber GB, Mattelin G, Lambiet-Collier M: Chemical protection against the long-term effects of a single whole-body exposure of mice to ionizing radiation. II. Causes of death. *Radian Res* 1978; 74:415-35.
98. McGinnis JE, Proctor PH, Demopoulos HB, Hokanson JA, Van NT: In vivo evidence for superoxide and peroxides production by adriamycin and cis-platium, Autor A (ed.) *Active Oxygen and Medicine.* New York, Raven Press, 1980.
99. Petkau A: Radiation protection by superoxide dismutase. *Photochem Photobiol* 1978;28:765-74.
100. Schaefer A, Komlos M, Seregi A: Lipid peroxidation as the cause of the ascorbic acid induced increase of ATPase activities of rat brain microsomes and its inhibition by biogenic amines and psychotropic drugs. *Biochem Pharmcol* 1975;24:1781-86.
101. Vladimirov YA, Sergeer PV, Seifulla RD, Rudnev YN: Effect of steroids on lipid peroxidation and liver mitochondrial membranes: *Molekuliarnaia Biologiia* (Russian, Moscow) 1973;7:247-62. (Translated by Consultants Bureau, a division of Plenum Publishing Inc., New York).
102. Sies H, Summer KH: Hydroperoxide-metabolizing systems in rat liver. *Eur J Biochem* 1975;57:503-12.
103. Alfthan G, Pikkarainen J, Huttunen JK, Puska P: Association between cardiovascular death and myocardial infarction and serum selenium in a matched-pair longitudinal study. *Lancet* 1982;2(8291):175-79.
104. Willett WC, Morris JS, Pressel S, *et al*: Prediagnostic serum selenium and risk of cancer. *Lancet* 1983;2(8343):130-34.
105. Demopoulos HB, Flamm ES, Seligman ML, Jorgensen E, Ransohoff J: Antioxidant effects of barbiturates in model membranes undergoing free radical damage. *Acta Neurol Scand* 1977;56(suppl 64):152.
106. Flamm ES, Demopoulos HB, Seligman ML, Mitamura JA, Ransohoff J: Barbiturates and free radicals, in Popp AJ, Popp RS, Bourke LR, Nelson, Kimelberg HK (eds.): *Neural Trauma.* New York, Raven Press, 1979, pp 289-300.

107. Butterfield DJ, McGraw CP: Free radical pathology. *Stroke* 1978;9(5):443-45.
108. Demopoulos HB, Flamm ES, Pietronigro DD, Seligman ML: The free radical pathology and the microcirculation in the major central nervous system disorders. *Acta Physiol Scand* 1980;492(suppl):91-119.
109. Morel DW, Hessler JR, Chisolm GM: Low density lipoprotein cytotoxicity induced by free radical peroxidation of lipid. *Journal of Lipid Research* 1983;24: 1070-76.
110. Smith LL: *Cholesterol Autooxidation*. New York, Plenum Press, 1981.
111. Gaby AR: Nutritional factors in cardiovascular disease. *J Holistic Med* 1983;5(2):815-828.
112. Taylor CB, Peng SK, Werthessen NT, Tham P, Lee KT: Spontaneously occurring angiotoxic derivatives of cholesterol. *Am J Clin Nutr*1979;32:40.
113. Passwater PA, Cranton EM: *Trace Elements. Hair Analysis and Nutrition.* New Canaan, Connecticut, Keats Publishing, Inc., 1983.
114. Fourcans B: Role of phospholipids in transport and enxymic reactions. *Adv Lipid Res* 1974;147-226.
115. Babior BM: Oxygen-dependent microbial killing by phagocytes. *N Engl J Med* 1978;298:659-68.
116. Rosen H, Klebanoff SJ: Bactericidal activity of a superoxide anion-generating system. *J Exp Med* 1979;149:27-39.
117. Masterson WL, Slowinski E: *In Chemical Principles*. Philadelphia, Saunders, 1977, p 203, plate 5.
118. Mayes PA: Biologic oxidation, In Martin DW, Mayes PA, Rodwell VW (eds): *Harper's Review of Biochemistry*. Los Alton, California, Lange Medical Publications, 1983;129-30.
119. March J: *Advanced Organic Chemistry Reactions, Mechanics, and Structure,* ed 2. New York, McGraw-Hill, 1978, 620.
120. Flamm ES, Demopoulos HB, Seligman ML, Poser RG, Ransohoff J: Free radicals in cerebral ischemia. *Stroke* 1978;9(5):445-47.
121. Hyperbaric Oxygen Therapy: A Committee Report, February 1981. Undersea Medical Society, Inc., 9650 Rockville Pike, Bethesda MD 20014 (UMS Publication Number 30 CR(HBO) 2-23-81).
122. Foote CS: Chemistry of singlet oxygen VII. Quenching by beta carotene. *J Am Chem Soc* 1968;90:6233.
123. Peto R, Doll R, Buckley JD, Sporn MG: Can dietary beta-carotene materially reduce human cancer rates? *Nature* 1981;290:201.
124. Schaefer A, Komlos M, Seregi A: Lipid peroxidation as the cause of the ascorbic acid induced decrease of ATPase activities of rat brain microsomes and its inhibition by biogenic amines and psychotropic drugs. *Biochem Pharmacol* 1975;24:1781-86.
125. Ito T, Allen N, Yashon D: A mitochondrial lesion in experimental spinal cord trauma. *J Neurosurg* 1978;48:434-42.
126. Athersclerosis and auto-oxidation of cholesterol. *Lancet* 1980;ii:964-65.
127. Pietronigro DD, Demopoulos HB, Hovesepian M, Flamm ES: Brain ascorbic acid (AA) depletion during cerebral ischemia. *Stroke* 1982;13(1):117.

128. Demopoulos HB, Flamm Es, Seligman ML, Pietronigro DD, Tomasula T, DeCrescito V: Further studies on free radical pathology in the major central nervous system disorders: effect of very high doses of methylprednisolone on the functional outcome; morphology, and chemistry of experimental spinal cord impact injury. *Can J Physiol Pharacol* 1982;60(11):1415-24.
129. Flamm ES, Demopoulos HB, Seligman ML, Poser RG, Ransohoff J: Free radicals in cerebral ischemia. *Stroke* 1978;9:445.
130. Demopoulos HB, Flamm ES, Seligman ML, Pietronigro Dd: Oxygen free radicals in central nervous system ischemia and trauma, in Autor AP (ed.): *Pathology of Oxygen*. New York, Academic Press, 1982;127-55.
131. Demopoulos HB, Flamm ES, Seligman ML, Ransohoff J: Molecular pathogenesis of spinal cord degeneration after traumatic injury, in Naftchi NE (ed.): *Spinal Cord Injury.* New York and London, Spectrum Productions, Inc. 1982;45-64.
132. Sukoff MH, Hollin SA, Espinosa OE, *et al.*: The protective effect of hyperbaric oxygenation in experimental cerebral edema. *J Neurosur* 1968;29:236-39.
133. Kelly DL, Jr., Lassiter KRL, Vongsvivut A, *et al.*: Effects of hyperbaric oxygen and tissue oxygen studies in experimental paraplegia. *J neurosurg* 1972;36: 425-29.
134. Holbach KH, Wassman H, Hoheluchter KL, *et al.*: Clinical course of spinal lesions treated with hyperbaric oxygen. *Acta Neurochir* 1975;31:297-98.
135. Holbach KH, Wassman H, Linke D: The use of hyperbaric oxygenation in the treatment of spinal cord lesions. *Eur Neurol* 1977;16:213-21.
136. Yeo JD, Stabback S., McKinsey B: Study of the effects of hyperbaric oxygenation on experimental spinal cord injury. *Med J Aust* 1977;2:145-147.
137. Jones RF, Unsworth IP, Marasszeky JE: Hyperbaric oxygen and acute spinal cord injuries in humans. *Med J Aust* 1978;2:573-75.
138. Yeo JD, Lawry C: Preliminary report on ten patients with spinal cord injuries treated with hyperbaric oxygenation. *Med J Aust* 1978;2:572-73.
139. Gelderd JB, Welch DW, Fife WP, *et al.*: Therapeutic effects of hyperbaric oxygen and dimethyl sulfoxide following spinal cord transections in rats. *Underse Biomedical Research* 1980;3:305-20.
140. Sukoff MH: Central nervous system: Review and update cerebral edema and spinal cord injuries. *HBO Review* 1980;1:189-95.
141. Jesus-Greenberg DA: Acute spinal cord injury and hyperbaric oxygen therapy: A new adjunct in management. *Journal of Neurosurgical Nursing* 1980;12:155-60.
142. Higgins AC, Pearlstein MS, Mullen JB, *et al.*: Effects of hyperbaric oxygen therapy on long-tract neuronal conduction in the acute phase of spinal cord injury. *J Neurosurg* 1981;55(4):501-10.
143. Sukoff MH, Ragatz RE: Use of hyperbaric oxygen for acute cerebral edema. *Neurosurgery* 1982;10:29-38.
144. De La Torre JC, Johnson CM, Goode DJ, Mullan S: Pharmacologic treatment and evaluation of permanent experimental spinal cord trauma. *Neurology* 1975;25: 508-14.
145. De La Torre JC, Kawanaga HM, Rowed DW, *et al.*: Dimethyl sulfoxide in central nervous system trauma. *Ann NY Acad Sci* 1975;243:362-89.

146. De La Torre JC, Surgeon JW: Dexamethasone and DMSO in experimental transorbital cerebral infarction. *Stroke* 1976;7:577-83.
147. Laha RK, Dujovny M., Barrionuevo PJ, *et al.*: Protective effects of methyl prednisolone and dimethyl sulfoxide in experimental middle cerebral artery embolectomy. *J Neurosurg* 1978;49:508-16.
148. Fischer BH, Marks M, Reich T: Hyperbaric-oxygen treatment of multiple sclerosis. *N Engl J Med* 1983;308:181-86.
149. Swank R: *A Biochemical Basis of Multiple Sclerosis*. Springfield, Illinois, Charles C. Thomas, 1961.
150. Cimino JA, Demopoulos HB: Introduction: Determinants of cancer relevant to prevention, in the war on cancer. *Journal of Environmental Pathology and Toxicology* 1980;3:1-10.
151. Seligman ML, Flamm ES, Goldstein BD, Poser RG, Demopoulos HB, Ransohoff J: Spectrofluorescent detection of malonaldehyde as a measure of lipid free radical damage in response to ethanol potentiation of spinal cord trauma. *Lipids* 1977;12(11):945-50.
152. Flamm ES, Demopoulos HB, Seligman ML, Tomasula JJ, DeCrescito V, Ransohoff J: Ethanol potentiation of central nervous system trauma. *J. Neurosurg* 1977;46: 328-34.
153. Dix T: Metabolism of polycyclic aromatic hydrocarbon derivatives to ultimate carcinogens during lipid peroxidation. *Science* 1983;221:77.
154. Samuelsson B: Leukotrienes: Mediators of immediate hypersensitivity reactions and inflammation. *Science* 1983;220:568-75.
155. Hess ML, Manson NH, Okabe E: Involvement of free radicals in the pathophysiology of ischemic heart disease. *Can J Physiol Pharmacol* 1982;60(11):1382-89.
156. Vincent GM, Anderson JL, Marshall HW: Coronary spasm producing coronary thrombosis and myocardial infarction. *N Engl J Med* 1983;309(14):220-39.
157. Harmon D: The free radical theory of aging. Read before the Orthomolecular Medical Society, San Francisco, California, May 8, 1983. (Available on audio cassette form AUDIO-STATS, 3221 Carter Avenue, Marina Del Rey, CA 90291).
158. Singal PK, Kapur N, Dhillon KS, Beamish RE, Dhalla NS: Role of free radicals in catecholamine-induced cardiomyopathy. *Can J Physiol Pharmacol* 1982;60(11):1340-97.
159. Crapper-McLaughlin DR: Aluminum Toxicity in senile dementia: Implications for treatment. Read before the Fall Conference, American Academy of Medical Preventics, Las Vegas, Nevada, Nov 8, 1981. (Available on audio cassette from AAMP, 6151 West Century Blvd., #1114, Los Angeles CA 90045).
160. Raymond JP, Merceron R, Isaac R, Wahbe F: Effects of EDTA and hypercalcemia on plasma prolactin, parathyroid hormone and calcitonin in normal and parathyroidectomized individuals. Read before the Frances and Anthony D'Anna International Memorial Symposium, Clinical Disorders of Bone and Mineral Metabolism, May 8, 1983. (Abstract available from Henry Ford Hospital, Dearborn, Michigan).
161. Frost HB: Coherence treatment of osteoporosis. *Orthop Clin North AM* 1981;12:649-69.

162. Deluca HF, Frost HM, Jee WSS, *et al.* (eds.): *Osteoporosis: Recent Advances in Pathogenesis and Treatment.* Baltimore, University Park Press, 1981.
163. Frost HM: Treatment of osteoporosis by manipulation of coherent bone cell populations. *Clin Orthop* 1979;143:224-44.
164. Meyer MS, Chalmers TM, Reynolds JJ: Inhibitory effect of follicular stimulating hormone in parathormone in rat calvaria in vitro. Read before the Frances and Anthony D'Anna International Memorial Symposium, Clinical Disorders of Bone and Mineral Metabolism, May 8, 1983. (Abstract available from Henry Ford Hospital, Dearborn, Michigan).
165. Wills ED: Lipid peroxide formation in microsomes. *Biochem J* 1969;113:325-32.
166. Gutteridge JMC, Rowley DA, Halliwell B, Westermarck T: Increased non-protein-bound iron and decreased protection against superoxide-radical damage in cerebrospinal fluid from patients with neuronal ceroid lipfuscinoses. *Lancet* 1982;ii:459-60.
167. Wilson RL: Iron, zinc, free radicals anjud oxygen tissue disorders and cancer control, in Iron Metabolism, *Ciba Found Symp 51* (new series). Amsterdam: Elsevier, 1977; 331-54.
168. Gutteridge, JMC: Fate of oxygen free radicals in extracellular fluid. *Biochem Soc Trans* 1982;10:72-74.
169. Wills ED: Mechanisms of lipid peroxide formation in tissues: Role of metals and haematin proteins in the catalysis of the oxidation of unsaturated fatty acids. *Biochem Biophys Acta* 1965;98:238-51.
170. Gutteridge JMC, Rowley DA, Halliwell B: Superoxide-dependent formation of hydroxyl radicals and lipid peroxidation in the presence of iron salts. *Biochem J* 1982;206:605-9.
171. Heys AD, Dormandy TL: Lipid peroxidation in iron-overloaded spleens. *Clinical Science* 1981;60:295-301.
172. Ericson JE, Shirahata H, Patterson CC: Skeletal concentrations of lead in ancient Peruvians. *N Engl J Med* 1979;300:946-51.
173. Schroder HA: *The Poisons Around Us.* Bloomington, Indiana, Indiana University Press, 1974;49.
174. Jenkins DW: *Toxic Trace Metals in Mammalian Hair and Nails.* US Environmental Protection Agency publication No. (EPA)-600/4-79049. Environmental Monitoring Systems Laboratory, 1979. (Available from National Technical Information Service, U.S. Department of Commerce, Springfield VA 22161).
175. Cranton EM, Bland JS, Chatt A, Krakovitz R, Wright JV: Standardization and interpretation of human hair for elemental concentrations. *J Holistic Med* 1982;4: 10-20.
176. Hansen JC, Christensen LB, Tarp U: Hair lead concentration in children with minimal cerebral dysfunction. *Danish Med Bull* 1980;27:259-62.
177. Medeiros DM, Pellum LK, Brown BJ: The association of selected hair minerals and anthropometric factors with blood pressure in a normosensitive adult population. *Nutr Research* 1983;3:51-60.
178. Moser PB, Krebs NK, Blyler E: Zinc hair concentrations and estimated zinc intakes of functionally delayed normal sized and small-for-age children. *Nutr Research* 1982;2:585-90.

179. Thimaya S, Ganapathy SN: Selenium in human hair in relation to age, diet, pathological condition and serum levels. *Sci Total Enviorn* 1982;24:41-49.
180. Musa-Alzubaida L, Lombeck I, Kasperek K, Feinendegen LE, Bremer HJ: Hair selernium content during infancy and childhood. *Eur J Pediatr* 1982;139:295-96.
181. Gibson RS, Gage L: Changes in hair arsenic levels in breast and bottle fed infants during the first year of infancy. *Sci Total Environ* 1982;26:33-40.
182. Ely DL, Mostardi RA, Woebkenberg N, Worstell D: Aerometric and hair trace metal content in learning-disabled children. *Environ Res* 1981;25(2):325-39.
183. Yokel RA: Hair as an indicator of excessive aluminum exposure. *Clin Chem* 1982;28(4):662-65.
184. Bhat RK, *et al.*: Trace elements in hair and environmental exposure I. *Sci Total Environ* 1982;22(2):169-78.
185. Hurry VJ, Gibson RS: The zinc, copper, and manganese status of children with malabsorption syndromes and inborn errors of metabolism. *Biol Trace Element Res*1982;4:157-73.
186. Thatcher RW, Lester ML, McAlester R, Horst R: Effects of low levels of cadmium and lead on cognitive functioning in children. *Arch Enviorn Health* 1982;37(3):159-66.
187. Peters HA, Croft WA, Woolson EA, Darcey BA: Arsenic, chromium, and copper poisoning from burning treated wood. *N Engl J Med* 1983;308(22):1360-61.
188. Yamanaka S, Tanaka H, Nishimura M: Exposure of Japanese dental workers to mercury. *Bull Tokyo Den Coll* 1982;23:15-24.
189. Capel ID, Spencer EP, Levin HN, Daivies AE: Assessment of zinc status by the zinc tolerance test in various groups of patients. *Clin Biochem* 1982;15(2): 257-60.
190. Vanderhoff JA, *et al.*:Hair and plasma zinc levels following exclusion of biliopancreatic secretions from functioning gastrointestinal tract in humans. *Dig Dis Sci* 1983;28(4):300-5.
191. Foli MR, Hennigan C, Errera J: A comparison of five toxic metals among rural and urban children. *Environ Pollut Ser A Ecol Biol* 1982;29:261-270.
192. Collipp PJ, Kuo B, Castro-Magana M, Chen SY, Salvatore S: Hair zinc levels in infants. *Clin Pediatr* 1983;22(7):512-13.
193. Medeiros DM, Borgman RF: Blood pressure in young adults as associated with dietary habits, body conformation, and hair element concentrations. *Nutr Res* 1982;2:455-66.
194. Huel G, Boudene C, Ibrahim MA: Cadmium and lead content of maternal and newborn hair: Relationship to parity, birth weight, and hypertension, *Arch Environ Health* 1981;35(5):221-27.
195. Marlowe M, Folio R, Hall D, Errera J: Increased lead burdens and trace mineral status in mentally retarded children. *J Spec Educ* 1982;16:87-99.
196. Marlowe M, Errera J, Stellern J, Beck D: Lead and mercury levels in emotionally disturbed children. *J Orthomol Psychiatr* 1983;12(4):260-67.
197. Nolan KR: Copper toxicity syndrome. *J Orthomol Psychiatr* 1983;12(4):270-82.
198. Klevay L: Hair as a biopsy material–assessment of copper nutriture. *Am J Clin Nutr* 1970;23(8):1194-202.

199. Rees EL: Aluminum poisoning Papua New Guinea natives as shown by hair testing. *J Orthomol Psychiatr 1983;12(4):312-13.*
200. Cook JD, Finch CA, Smith NJ: Evaluation of the iron status of a population. *Blood* 1976;48:449-55.
201. Kannel WB, Hjortland MC, McNamara PM, Gordon T: Menopause and the risk of cardiovascular disease. The Framingham Study. *Ann Intern Med* 1976;85: 447-52.
202. Hjortland MC, McNamara PM, Kannel WB: Some atherogenic concomitants of menopause: The Framingham Study. *Am J Emidemiol* 1976;103:304-111.
203. Gordon T, Kannel WB, Hjortland MC, McNamara PM: Menopause and coronary heart disease. The Framingham Study. *Ann Intern Med* 1978;89:157-61.
204. Sullivan JL: Iron and the sex difference in heart disease risk. *Lancet* 1981;1(8233):1293-94.
205. Skoog DA, West DM: Volumetric methods based on complex-formation reactions, in *Fundamentals of Analytical Chemistry,* New York, Holt, Rinehart and Winston, Inc., 1969;338-60
206. Peng CF, Kane JJ, Murphy ML, Straub KD: Abnormal mitochondrial oxidative phosphorylation of ischemic myocardium reversed by calcium chelating agents. *J Mol Cel Cardiol* 1977;9:897-908.
207. Freeman AP, Giles RW, Berdoukas VA, Walsh WF, Choy D, Murray PC: Early left ventricular dysfunction and chelation therapy in thalassemia major. *Ann Intern Med 1983;99:450-54.*
208. Blake DR, Hall ND, Bacon PA, Dieppe PA, Halliwell B, Gutteridge JMC: Effect of a specific iron chelating agent on animal models of inflammation. *Ann Rheum Dis* 1983;99:450-54.
209. Addonizo VP, Wetstein L, Fisher CA, Feldman P, Strauss JF, Harken AH: Medication of cardiac ischemia by thromboxanes released from human platelets. *Surgery* 1982;92:292.
210. Yagi K, Ohkawa H, Ohishi N, Yamashita M, Nakashima T: Lesion of aortic intima caused by intravenous administration of linoleic acid hyperoxide. *J Appl Biochem* 1981;3:58-65.
211. Benditt EP: The origin of atherosclerosis. *Scientific American* Feb 1977; pp 74-85.
212. McCullach KG: Revised concepts of atherogenesis. *Cleve Clin Q* 1976;43:247.
213. Mayron LW: Portals of entry–a review. *Ann Allerg* 1978;40:399-405.
214. Walter WA, Isselbacher KJ: Uptake and transport of macromolecules by the intestine: Possible role in clinical disorders. *Gastroenterology* 1974;67:631-50.
215. Hemmings WA, Williams EW: Transport of large breakdown products of dietary protein through the gut wall. *Gut* 1978;19:715-23.
216. Rowe AH, Rowe AH Jr: *Food Allergy; Its Manifestations and Control and the Elimination Diets.* Springfield, Illinois, Charles C. Thomas, 1972.
217. Speer F (ed.): *Allergy of the Nervous System.* Springfield, Illinois, Charles C. Thomas, 1970.
218. Dickey LD (ed.): *Clinical Ecology.* Springfield, Illinois, Charles C. Thomas, 1976.

219. Randolph TG: *Human Ecology and Susceptibility to the Chemical Environment.* Springfield, Illinois, Charles C. Thomas, 1962.
220. Crook WG: The coming revolution in medicine. *J Tenn Med Assn* 1983;76(3): 145-49.
221. Iwata K: Toxins produced by Candida albicans. *Contr Microbiology Immunol* 1977;4:77-85.
222. Iwata K, Yamamota Y: Glycoprotein toxins produced by Candida albicans. Reprinted from Proceedings of the Fourth International Conference on the Mycosis. *PAHO Scientific Publication* 1977;356:246-57.
223. Iwata K: Fungal toxins as a parasitic factor responsible for the establishment of fungal infections. *Mycopathologia.* 65:141-54.
224. Crook WG: The Yeast Connection: *A Medical Breakthrough.* Jackson, TN, Professional Books, 1983 (P.O. Box 3494, Jackson, TN 38301).
225. Truss CO: Tissue injury induced by Candida albicans. *Orthomolecular Psychiatry* 1978;7(1):17-37.
226. Truss CO: Restoration of immunologic competence to Candida albicans. *Orthomolecular Psychiatry* 1980;9(4):287-301.
227. Truss CO: The Role of Candida albicans in human illness. *Orthomolecular Psychiatry* 1980;9(4):228-38.
228. Hatano S, Nishi Y, Usui T: Copper levels in plasma and erythrocytes in healthy Japanese children and adults. *Am J Clin Nutr* 1982;35:120-26.
229. Harrison W, Yuracheck J, Benson C: The determination of trace elements in human hair by atomic absorption spectroscopy. *Clin Chim Acta* 1969;23(1): 83-91.
230. Gori GB: Observed no-effect thresholds and the definition of less hazardous cigarettes. *Journal of Environmental Pathology and Toxicology* 1980;3:193-203.
231. Saltin B, Karlsson J: *Muscle Metabolism During Exercise.* New York, Plenum Publishing Co., 1971; p 395.
232. de Jesus-Greenberg D: Hyperbaric oxygen therapy. *Critical Care Update* 1981;8(2): 8-20.
233. Important Facts You Should Know About the American Board of Chelation Therapy, 1983. American Board of Chelation Therapy, 1312 N.W. Twelfth Street, Moore, Oklahoma 73170.
234. American Board of Chelation Therapy. Constitution and By-Laws, April 1983. American Board of Chelation Therapy, 1312 N.W. Twelfth Street, Moore, Oklahoma 73170.
235. Protocol for the Safe and Effective Use of EDTA Chelation Therapy, 1983. American Board of Chelation Therapy, 1312 N.W. Twelfth Street, Moore, Oklahoma 73170.